Human Behavior and Environment

ADVANCES IN THEORY AND RESEARCH

Volume 1

Human Behavior and Environment

ADVANCES IN THEORY AND RESEARCH

Volume 1

EDITED BY

IRWIN ALTMAN

University of Utah
Salt Lake City, Utah

AND

JOACHIM F. WOHLWILL

The Pennsylvania State University
University Park, Pennsylvania

PLENUM PRESS · NEW YORK AND LONDON

© 1976 Plenum Press, New York
A Division of Plenum Publishing Corporation
227 West 17th Street, New York, N.Y. 10011

Printed in the United States of America

Contributors

DAVID C. MERCER · Department of Geography, Monash University, Melbourne, Australia

TIMOTHY O'RIORDAN · School of Environmental Sciences, University of Anglia, England

H. McILVAINE PARSONS · Institute for Behavioral Research, Incorporated, Silver Spring, Maryland

KERMIT K. SCHOOLER · School of Social Work, Syracuse University, Syracuse, New York

EDWIN P. WILLEMS · Department of Psychology, University of Houston, Houston, Texas

JOACHIM F. WOHLWILL · Division of Man–Environment Relations, The Pennsylvania State University, University Park, Pennsylvania

ERWIN H. ZUBE · Institute for Man and Environment, University of Massachusetts, Amherst, Massachusetts

Preface

This is the first in a series of volumes concerned with research encompassed by the rather broad term "environment and behavior." The goal of the series is to begin the process of integration of knowledge on environmental and behavioral topics so that researchers and professionals can have material from diverse sources accessible in a single publication.

The field of environment and behavior is broad and interdisciplinary, with researchers drawn from a variety of traditional disciplines such as psychology, sociology, anthropology, geography, and other social and behavioral sciences, and from the biological and life sciences of medicine, psychiatry, biology, and ethology. The interdisciplinary quality of the field is also reflected in the extensive involvement of environmental professionals from architecture, urban planning, landscape architecture, interior design, and other fields such as recreation and natural resources, to name just a few.

At present, the field has a somewhat chaotic flavor, with research being carried out by a variety of scholars who publish in a multitude of outlets. Many researchers and practitioners are unaware of the state of knowledge regarding a specific topic because of the unavailability of integrated reference materials. There are only a handful of books dealing with environment and behavior, most of them unintegrated collections of readings, with only an occasional systematic analysis of some facet of the field. The present series of volumes is based on the assumption that research has proceeded to the point where it is useful to initiate the process of integration of knowledge about selected environmental and behavioral topics. Furthermore, we believe that the time is *not* yet ripe for a single compendium, a "handbook" of research on environmental and behavioral topics. Rather, a series of volumes with each reviewing selected topics can provide an evolutionary accrual of knowledge in this burgeoning field. To this end we have solicited originally written chapters on a range of current research topics in the environmental and behavioral field, and we plan a two-volume sequence that will collectively sample a broad range of issues. Authors were invited to contribute chapters tailored to the issues they were to discuss in terms of well-established principles, gaps, inconsistencies, and directions for future research. They were also asked to point to crucial research questions on their topics, and to provide a personal perspective deriving from their own research.

This first volume covers such topics as environmental attitudes, environmental aesthetics, perceptual aspects of land use, environmental change and the aged, ecological psychology applied to institutional settings, recreation, and work environments. The second volume in the series will continue this survey of the overall field of environment and behavior, treating such topics as personal space, energy, environmental stress, evaluation of environmental design, simulation, and operant approaches to environment and behavior, among others.

The intent in the first two volumes is to sample broadly, so as to illustrate the range of research knowledge on various topics in the field of environment and behavior. There will necessarily be some disconnectedness between chapters because of the different integrative approaches taken by authors, as well as the deliberate absence of topically defined communalities between them.

Beginning with the third volume in the series, some two years hence, a thematic approach will be taken, with individual volumes built around specific field environmental and behavioral topics. By that time, we believe that the field will have sufficiently matured so that researchers and practitioners will profit by having subject matter on a single topic critically analyzed and integrated in a single volume.

As we see it, then, successive volumes in this series will reflect the evolutionary development of research on environment and behavior. As subareas of the field grow and warrant systematization, we plan thematic volumes devoted to these subareas. Yet, even in the face of such increasing specialization, we hope to maintain a strong integrative focus in the contributions to this series in a dual sense: We will aim to solicit papers integrating fact and theory from authors able to and interested in undertaking such a task; we hope likewise to encourage the integration of material from diverse disciplines bearing on a particular problem or theory, so that this series will help to bring together the rather scattered persons, ideas, and findings that characterize the field of environment and behavior today.

Irwin Altman
Joachim F. Wohlwill

Contents

CHAPTER 3

PERCEPTION OF LANDSCAPE AND LAND USE

ERVIN H. ZUBE

CHAPTER 4

MOTIVATIONAL AND SOCIAL ASPECTS OF RECREATIONAL BEHAVIOR

DAVID C. MERCER

CHAPTER 5

WORK ENVIRONMENTS

H. MCILVAINE PARSONS

CHAPTER 6

BEHAVIORAL ECOLOGY, HEALTH STATUS, AND HEALTH
CARE: APPLICATIONS TO THE REHABILITATION SETTING

EDWIN P. WILLEMS

CHAPTER 7

ENVIRONMENTAL CHANGE AND THE ELDERLY

KERMIT K. SCHOOLER

Attitudes, Behavior, and Environmental Policy Issues

TIMOTHY O'RIORDAN

How many Americans, I wonder, remember the long lineups to obtain gas in the latter part of January 1974? The impact of that national crisis is a distinct memory for some, perhaps, but probably not for the majority. And what about the effect of such recollections on current energy consumption? Surprisingly, little, even though prices have doubled for all kinds of energy compared with two years ago. For example, gasoline consumption has remained steady, almost oblivious to the 50% increase in prices since 1973, and while thermostats were lowered by 2°F or even 3°F during the winter of 1974, this practice did not persist. In any case it reduced domestic fuel demand by less than 2% since substantial heat losses still occurred due to inadequate insulation, a legacy of the era of cheap fuel and improperly scrutinized building codes (Ford Energy Policy Project, 1974, pp. 119–120).

Although most Americans realize that some sort of "energy problem" exists, there is as yet little evidence that this knowledge has been converted into substantial and sustained energy-conserving behavior, although most experts stress that the best way out of the energy dilemma is substantially to reduce demand (ibid.). This leads to the question, What is the relationship between knowledge, attitudes, and behavior with respect to environmental policy issues? What guidelines should policymakers and educators follow in order to

TIMOTHY O'RIORDAN · School of Environmental Sciences, University of East Anglia, England.

obtain voluntary and effective public response? The aim of this chapter is to explore this relationship, both theoretically and empirically, in an effort to understand the political issues involved.

Murray and his colleagues (1974) at the National Opinion Research Center provide some revealing information that gives us some clues to the tortuous relationships between environmental attitudes and behavior. As part of a continuous national survey, they interviewed some 600 Americans every week throughout the period December 1973 to February 1974, during the time of the Middle East oil embargo and the quintupling of world oil prices. They discovered an interesting cycle of attitudes and behavior as the grip of the "energy crisis" tightened and then apparently relaxed. To begin with, a plurality responded to a presidential request to reduce oil consumption. Not surprisingly, this response came mainly from the more affluent, who in any case consume disproportionately large quantities of energy; they can cut back without much discomfort, but in aggregate save a sizable amount of fuel. The poorer sectors of society, who use up less energy per capita, have less room for maneuver, but in any case a reduction in their demand has little impact on total energy consumption. It is a moot point whether the rich are more socially motivated than the poor during a period of "national emergency" and thus feel a stronger social obligation to change their behavior. The evidence points this way because there is a positive connection between education and social concern and, of course, between education and income, but the influence of such public-spiritedness on sustained behavioral response is still largely conjectural.

As the gasoline shortage became more evident, Murray and his associates found a distinct shift in public attitudes toward energy use and appropriate public policy. Their data showed a strong correlation with income and difficulty in obtaining gasoline. As noted above, there was a positive relationship between efforts to conserve energy and income, but a much stronger relationship existed between persistent difficulty in getting fuel, conserving behavior, and opinions about the severity of the situation. The more direct the impact of the fuel shortage on personal habits, the greater was the pessimism about America's ability to surmount the energy problem, and the greater was the willingness to accept rationing. In addition, there was an increased awareness about energy usage generally and a stronger belief that the individual could do something about the whole "energy problem."

The Murray team also reported the amazing volatility of public opinion in response to news stories. During the third week of January, 1974, for instance, a number of the major oil companies reported vast increases in profits. Immediately blame for the "crisis" shifted away from the government (the usual scapegoat) to the oil companies (the more obvious target). At the same time truckers were threatening to strike in an attempt to reduce fuel prices and raise speed limits, so public opinion shifted to place trucks high in the list of priorities should there be widespread gasoline rationing.

But regardless of all the fuss, the public's attitudes and behavior toward the whole energy question did not appear to change significantly. Automobile use, especially for commuting (where most gasoline is consumed, anyway), hardly altered. There was a tremendous resistance to forming car pools, even for the short duration of the embargo, and even less willingness to switch to public transport (where it existed). Most important, as soon as gasoline became more readily available (albeit at a higher price) following the political solution to the Yom Kippur war, attitudes toward rationing, energy conservation, pricing, and other policy issues all but reverted to their pre-December state. Relieved Americans quickly came to believe the whole matter was transitory, an issue largely manipulated by politics and the power of the multinational oil companies, and that given the political determination, the nation could be energy independent within ten years.

Subsequent political events have largely been based on interpretation of these views. The president and Congress have been locked in a long battle of wills over whether to raise domestic oil prices, and tax imported oil, thereby rationing use by the traditional market mechanism of price (this would mean an additional 40% increase in price over two years), while providing financial incentives for the oil industry to develop indigenous energy supplies; or whether to maintain domestic oil price controls and restrain use either by rationing or by punitive taxation (the fruits of which would be shared by all Americans). The issue, of course, is enormously complicated due to wider political considerations, but the point to stress here is the recognition by the president and Congress of the unpopularity of what appear to the public to be "unnecessary" rises in energy prices or "unacceptable" impositions on their normal pattern of living.

This is a deep-seated cultural phenomenon. The American people enjoy a life-style and an economic system based on abundant and cheap energy, drawn either from easily won indigenous sources (where the adverse environmental impact was local and politically "acceptable") or from foreign suppliers who were economically exploited over a long period of time (Otaiba, 1975). In real terms, oil and natural gas prices have *declined* since 1949 in the sense that, until the late sixties anyway, productivity gains and associated income rises outstripped the price rise of petroleum-based fossil fuels (Landsberg, 1973, p. 35; Nordhaus, 1974, p. 23). Many people find it difficult to believe that this era has ended and that America may not be able to provide abundant wealth and technological inventiveness to sustain continued economic growth into the indefinite future. This view is reinforced by the political rhetoric of a president seeking continued popularity when he proposed the establishment of a new Energy Independence Authority to finance the expansion of domestic energy supplies. President Ford proclaimed that the new authority

> . . . can stimulate economic growth. It can create new jobs. It can give us control over our own destiny. It can end runaway energy prices imposed by foreign nations. It can give foreign nations a new look at what Americans can do with our great resources when we stop talking and start acting.

Even at the depth of the "crisis" Murray and his associates found that less than a third of those interviewed regarded the energy issue as the most serious national problem, and less than a quarter felt there would be any kind of shortage after five years. They concluded:

> Short term expectations are influenced by exposure to shortages and themselves determine evaluations and conservation behaviors. Longer term ones are insensitive to recent experiences of shortages, relatively stable over time, and unrelated to evaluations of the energy shortage and conservation behaviors. (p. 259)

SOME PRELIMINARY OBSERVATIONS

The energy case study is cited because it points out some of the complicated relationships that exist between attitudes, behavior, and appropriate public policy with regard to environmental issues. It shows, for example, that there is enormous cultural inertia over changing accustomed beliefs and patterns of doing things, even given stringent behavioral limitations. It also demonstrates the difficulty facing policy advisers (and to a lesser extent, politicians) of persuading the public that there really is a need to change some of their cherished ways of life in order to avoid a major crisis of scarcity or environmental disruption. Many economists of the right (e.g., Ridker, 1973, pp. 31–32), the center (e.g., Galbraith, 1973), and the left (e.g., Weisberg,

1971) now believe that the price mechanism is so manipulated by government intervention and oligopolistic collusion that it simply cannot signal resource shortage or geopolitical maneuverings sufficiently far in advance of real scarcity. In any case, there is some doubt as to how far people can really respond effectively to price rationing during this period of high inflation and widespread social expectation that living standards will be maintained. The case study also highlights the problems confronting politicians of how to legitimize or justify government intervention in public behavior (for example, by restricting the use of automobiles so as to reduce gasoline consumption or air pollution) in the absence of a serious national emergency. Pirages and Ehrlich (1974, p. 62) observe that despite an appeal by the San Francisco Bay Area Pollution Control District to people to avoid using their cars whenever possible during a particularly dangerous smog episode, there was no demonstrable reduction in traffic flows. Winn (1973) reports that Angelenos have learned to laugh at the smog: It is part of their environment, it is an "imageable" aspect of their city, it is something to talk about, it is almost a "cult." Certainly few believe that it actually affects *their* health.

Finally the energy study shows up the seemingly impossible confusion in the relationship between information and opinion and behavior change in influencing attitudes and responsiveness to public policy. In attempting to come to grips with these linkages, social psychologists have traditionally experimented with voluntarily selected individuals in a controlled laboratory environment. Although conceptually comfortable, such paper and pencil tests fail to uncover the deeper cultural legacy of man's attitudes and behavior toward his natural surroundings and thus its relationships to the particular social-political institutions of Western societies. Such an analysis is fundamental if we are ever to come to grips with the complex phenomena of environmental words and environmental deeds, and is attempted in the section that follows.

MAN AND NATURE—AMBIVALENCE AND INCONSISTENCY

Man distinguishes himself from other living organisms in being able to conceptualize and order environmental phenomena into a coherent pattern of images, expectations, and meanings. All human cultures characterize their surroundings into complicated symbolic patterns in order to cope with the world and to come to terms with it. Yi Fu Tuan (1971, 1972, 1974) is probably the foremost scholar of the processes of environmental cognition by various human cultures. "Man is egocentric," he writes (1971, p. 4), "the word he organizes is centered upon himself." But he adds that man is also ethnocentric:

> Worlds, whether of individuals or of cultures are made up of the perceived elements in nature or external reality: they are distorted by human needs or desires; they are fantasies. It is a paradox that human beings can live in fantasy yet not only survive but prosper. Fantasy is more than gratuitous daydreaming: it is also man's effort to explain, to introduce order to life situations that so often seem baffling and contradictory.

Tuan (1971, p. 18) emphasizes that the act of cognizing environments has two purposes: One is to relate man to his natural surroundings, the other to help him structure his social world. Indeed, the relationships between social structure, economic organization, and environmental cognition are closely intertwined. Burch (1972, pp. 49–50) speculates that ephemeral nomadic tribes with dispersed subsistence economies and small populations held close to a steady state, developed rather simple patterns of environmental cognition, and evolved decentralized and largely egalitarian social structures. Sedentary agricultural communities, on the other hand, enjoyed production surpluses that led to the accumulation of wealth, more elaborate and centralized social relationships, and a much more sophisticated conceptualization of nature.

But regardless of the relationship between environmental cognition and socioeconomic organization, there is evidence that all human cultures display a fundamental inconsistency between an expressed (or written) environmental ethos and environmental behavior. This duality was first fully reviewed by Glacken (1967) and thoroughly examined by Tuan (1968, 1970, 1972). From his impressive study of man–nature attitudes from the Greeks to the end of the 18th century Glacken concluded that three contradictory beliefs dominated: (a) a recognition that man's actions are to some extent determined by his physical surroundings, (b) a knowledge that man is capable of causing ecological damage, and (c) a belief that the earth was designed by God for him to use so as to improve his mind as well as his economic and social conditions. Civilization was therefore regarded as the application of purposive order by which a pliable earth was molded to satisfy man's needs. Mastery over nature, if not by technology and artifact, at least by conceptualization and symbolism, was always a driving element in environmental construing.

But paradoxically mastery could never be so dominant as to threaten man's existence; hence the continuous role of symbolism to resolve the inconsistency. For example, Tuan (1971, p. 32) quotes the conflict evident in the minds of the Iglulik Eskimos who must kill to survive but never to grow too fat. As the Eskimos put it:

> The greatest peril of life lies in the fact that human food consists entirely of souls. All the creatures that we have to kill and eat, all those that we have to strike down and destroy to make clothes for ourselves, have souls, like we have, souls that do not perish with the body and which must therefore be propitiated lest they should avenge themselves on us for taking away their bodies.

There remains much intellectual and emotional controversy about the environmental behavior of prehistoric peoples and modern-day preindustrial societies. Some anthologies, such as *Touch the Earth* by McLuhan (1972) expound the "I–thou" view, namely, that these groups exhibited a thoughtfulness and care of the earth to the point of reverence that all but curtailed any threat of environmental deterioration. Other writers, like Martin (1967), Tuan (1970), and Guthrie (1971), take the view that all tribes inadvertently or deliberately exploited the earth and so were the agents of widespread ecological change. Martin talks of "prehistoric overkill," the extinction of whole species caused by Paleolithic hunting technologies. Guthrie is even more damning. "The Indian's attitudes toward nature," he comments (p. 722), "were and are identical to those of modern man. What concern for the environment there was existed for the express purpose of guaranteeing human survival." Tuan (1970) takes a more sensible middle view of this issue. "A culture's published ethos about its environment," he observes (1970, p. 244), "seldom covers more than a fraction of the total range of its attributes and practices pertaining to that environment. In the play of forces that govern the world, aesthetic and religious ideas rarely have a major role."

Tuan stresses that the inconsistency between words and deeds is culturally ingrained in homo sapiens precisely because all people can separate ideal from reality, moral responsibility from actual practice, blame from action. Because he is anthropocentric and egocentric, man does disassociate himself from the forces of nature and strives to create some of his own destiny for himself. Just as the U.S. Corps of Engineers can destroy sizable sections of the nation's land and waters for dams and harbors, because "with the country growing the way it is, we cannot simply sit back and let nature take its course" (quoted in Drew, 1970, p. 52), so the ancient Chinese cut and burned down large areas of forest to destroy the habitats of unfriendly animals and people, to clear the land with the minimum of effort, to utilize the soil, to provide fuel for warmth and for their cremations, and to build huge halls and temples (Tuan, 1970, p. 248).

While there is evidence that all cultures damage their habitats and seemingly recognize that they are doing so, it should be emphasized that all social behavior is subject to certain collectively sanctioned restraints that themselves are a function of social organization and economic institutions. Burch (1972, pp. 49–50) speculates that environmental exploitation occurs when the social order is in a high state of flux, when traditional social rules are no longer operable. This would occur with an alteration in the environmental energy budget (a change in climate) or through contact with a new culture and new technologies. (One can speculate that a modern postindustrialized society experiencing continuous "future shock" suffers a similar disorientation flux and lack of effective ethical sanctions.) Thus, the famous Tsembaga pig-killing rituals (Rappoport, 1967), hailed by anthropologists and ecologists as evidence that man can consciously limit depredation of the environmental "commons" and maintain harmonious relations with his warring neighbors, only occurred because the Tsembaga had no social or economic need for large numbers of pigs that otherwise would have foraged in the forest, thereby damaging their cropland. But should the pigs ever have become a source of wealth (say, in intertribal trade or for export to meat-eating Western cultures), it is quite probable that the Tsembaga ritual would have been replaced by environmental exploitation.

The role of social and economic institutions in shaping environmental behavior is also evident in the so-called Judeo-Christian debate. This was sparked off by an essay by the historian Lynn White, Jr. (1967), in which he theorized that the "Judeo-Christian ethic" was the cause of Western man's alienation from nature, his objectification of natural objects and processes, his anthropomorphism, and his quest for progress. "Christianity," he commented, (p. 1205), "is the most anthropocentric religion the world has seen . . . It has not only established a dualism of man and nature but has also insisted that it is God's will that man exploit nature for his proper ends." White concluded that the marriage of this Christian ethos with the revolutions in science and technology have produced our present-day "ecologic crisis."

The biologist John Black (1970) disputes this interpretation. He believes that the controversial divine commands to Adam and Eve to "be fruitful, and multiply, and replenish the earth and subdue it: and have dominion . . . over every living thing that moveth upon the earth," probably do not represent the earliest attitudes of man to his environment. He claims that they are post hoc rationalizations, by a relatively advanced technological civilization, of prophesies interpreted as commands that turned out as facts. He also points to the potential dichotomy between the interpretation of "subdue" and "replenish." To subdue is to exert force and command complete control, an exhortation to continue the struggle against an inhospitable land. But having "won" the struggle, man's obligation both to himself, and his divine Master, is to tend and care for this land so that it remains continuously productive. Van Arkel (1974, p. 27) supports Black's view. Most early civilizations (with the notable exception of the Greeks) regarded human toil as a necessary evil, the consequence of man's Fall, but also a liberating means to impose man's will upon the landscape. This dual concept of labor combined with the *imago dei* belief that man was master of all creation led naturally to an environmental attitude that the earth was a usufruct.

Throughout the Greek, Hebrew, and Roman periods, however, the check to excessive exploitation was the notion of stewardship, the middle path between environmental dominance and subservience. Black (1970, pp. 44–58) notes that in biblical times, the steward was a most important figure in the social and economic hierarchy. His duties were to manage his master's estate for profit while ensuring its long-term viability. In a symbolic sense, the steward was God's deputy on earth (namely, man) who applied intelligence, reason, and moral responsibility in his care of the sacred garden. This idea is still prevalent among the Amish and Pennsylvania "Dutch" cultures and is evident in their meticulous landscapes and abstemious way of life.

But for most Western people, this crucial notion of stewardship has all but disappeared. Why? Leiss (1972) attributes its demise to the rise of capitalism and colonialism, where the exploitation of the earth led in turn to the exploitation of other human beings. Approvingly (p. 195) he quotes C. S. Lewis: "Man's power over nature turns out to be a power exercised by some men over other men with nature as its instrument." Thus he contends that domination of nature is not so much an object in itself as a reflection of social organization and political philosophy. Leiss takes the Marxist line that capitalism promotes greed, selfishness, and a desire for acquisition without limit, attributes that can only exist through exploitation. So capitalism creates conflict and division between workers, between workers and management, between urban man and rural man and inevitably between man and nature. As a result, moral restraint is weakened with no socially sanctioned mode of conduct to replace it. Domination of nature and the lore of technology, he asserts, are not innate human traits nor even the products of a particular religious ethos, but the outcomes of a certain mode of social and political existence. If this is the case, then it is only possible to envisage a constructive change in environmental construing through "a new set of social institutions in which responsibility and authority are distributed widely among the citizenry and in which all individuals are encouraged to develop their critical faculties" (Leiss, 1972, p. 197).

It would appear then, that in "macropsychological" terms, attitudes and behavior toward the environment are in part a product of a lengthy cultural legacy, combined with a particular mode of economic production. But it is not a one-way influence, since man–nature attitudes in turn help to create social organization and thus influence behavior. Most probably there is a transactional relationship at work between all these factors that mutually shape each other as a culture evolves. If so, it will not be easy to reform environmental attitudes and behavior by partial and cosmetic remedies such as appealing to the public conscience or raising prices, as the energy case study clearly demonstrates. The issue is enormously more complicated than this and lies at the very heart of the way that we individually and collectively organize and view our world. Let us now turn to the "micropsychology" of personal attitude–behavior studies to unravel this relationship a little further.

ENVIRONMENTAL WORDS AND ENVIRONMENTAL DEEDS

For over forty years social psychologists have been worrying over the nature of the links (if indeed they exist) between verbal expressions of attitude and actual behavior (see Deutscher, 1966; Kiesler, Collins, and Miller, 1969; Wicker, 1969; Thomas, 1971). Generally speaking, the prevailing paradigm assumed a causal connection between words and deeds, first without any intervening variable, then with various overt and latent personal and situational factors mediating between attitude and behavior, and later still with an added component for social obligation (Figures 1a, 1b, 1c). Despite the rather cumbersome relationships evident in the later models, there remained a tacit acceptance, among most workers, of a unidirectional relationship between attitude and subsequent response executed by a rational and purposeful individual. For example, de Fleur and Westie (1963, p. 21) concluded:

> The attitude, then, is not the manifest responses themselves, or their probability, but an intervening variable operating between stimulus and response and inferred from overt behavior. This inner process is seen as giving both direction and consistency to the person's responses.

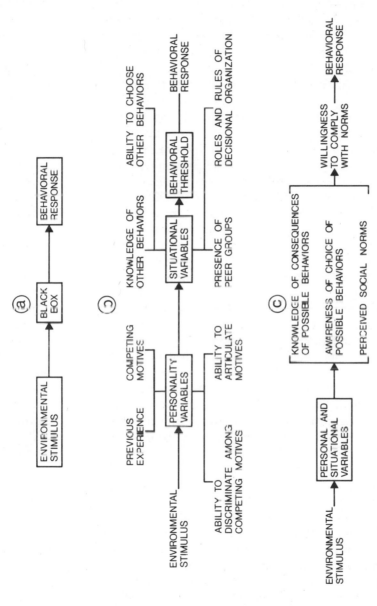

Figure 1. Three paradigms for the conceptualization of relationships between environmental stimuli and overt responses with reference to environmental attitudes.

Campbell (1963) visualizes attitudes as socially acquired behavioral dispositions, modified by experience, that guide behavior over a specific range of circumstances. Although he acknowledges that behavior partly determines experience, he concludes that attitudinal dispositions are the dominant force in guiding behavior. The prevailing belief generally holds firm that, provided attitudes are measured carefully, likely behavioral responses can be speculated upon with some degree of assurance. Despite the almost nihilistic declarations of some psychologists, such as Tarter (1970) who wrote that "attitudes as presently conceptualized play no real role in behavior," this plausible, but deceptively dangerous assumption still seems to be accepted by many policymakers and opinion pollsters.

As already noted, the attitude → behavior evidence was derived largely from "paper and pencil" tests conducted on psychology students under controlled laboratory conditions, based on an implicit attitude → behavior methodological paradigm. Field research has failed to substantiate these findings. From an exhaustive analysis of the literature Wicker (1969, p. 65), concludes that "taken as a whole, these studies suggest that it is considerably more likely that attitudes will be unrelated or only slightly related to overt behaviors than that attitudes will be closely related to actions . . . Only rarely can as much as 10% of the variance in overt behavioral measures be accounted for by attitudinal data." Either the models are imperfect or the methodologies are inadequate. Both Crespi (1972) and Lauer (1971) believe that the latter is the main cause of the problem. Lauer (p. 248) notes that "the fault lies both in the failure to create research designs that reflect the complexity of the problem and in the tendency to reject the importance of the proximate causes of overt behavior."

As to the problem of the model, Wicker (1969, p. 67–75) proposes a more comprehensive schema (Figure 1b) in which he postulates that a number of important "latent process variables" have to be considered. These mediating variables fall into two principal groups, namely, personal variables and situational factors. Items under the "personal variables" include other attitudes held by the individual, competing motives, verbal and intellectual aptitude to conceptualize and describe attitudes, and feelings of personal efficacy. "Situational variables" include the presence or absence of peer groups, authority roles, and sanctions relating to the social or organizational environment and the opportunity to choose different courses of action. Turning to the personal variables first, the matters of multiple cognitions and competing motives demand further scrutiny. Since there seems to be no commonly accepted definition of the term "cognition" it is used here to characterize discrete bundles of closely connected beliefs and feelings that relate to various facets of an object or situation, and that in combination constitute the mediating mechanism regarded by Wicker as an attitude.

It is helpful to visualize the relationship between these cognitive "bundles" in terms of a mental construction of significance as depicted in Figure 2. At the "core" of any attitude are the dominant (or salient) cognitions, closely connected to a person's value system, which are held strongly and cherished. Rokeach (1968, p. 164) believes that these will be functionally connected to a subject of instrumental and terminal values that need not necessarily be mutually consistent; hence the potential for multiple incompatibility among the cognitive clusters. These control cognitions will bear upon a whole range of experiences of which the situation or object under observation is but one.

Surrounding the dominant cognitions are more peripheral (or nonsalient) cognitions, less strongly connected to values, which may readily be exchanged for other equally peripheral cognitions. Quite probably, many of what can be called "socially acceptable" cognitions fall into this latter category, cognitions that lie below the awareness threshold, but are elicited by public opinion polls and questionnaires or during conversation with social peers. Most of the time they may have little direct bearing on behavior, but in certain circumstances, they may be quite significant. Kiesler *et al.* (1969, p. 35) sum up the methodological problems thus caused as follows:

> Attitudinal measures frequently make broad and philosophical attitudes salient, whereas behavioral measures make specific and immediately personal attitudes salient. We can therefore expect a low correlation between attitude and behavior in those cases where one set of subattitudes [cognitions] is salient to the testing situation, whereas another set [of cognitions] is salient in the behavioral situation.

There is always a danger, therefore, of picking up irrelevant cognitions and assuming that they will bear upon subsequent behavior. This is probably the most difficult problem facing researchers in environmental cognition, for not only are many of the peripheral cognitions not terribly critical, they may be inconsistent with other more relevant ones, because of the conflicting motives of the individual studied. Since many environmental objectives are apparently irreconcilable with other valued ends (jobs versus amenity, for example), the incompatibility of subattitudinal cognitions can add to the distortion between verbal statement and actual deed.

In an unpublished study of the environmental attitude–behavior

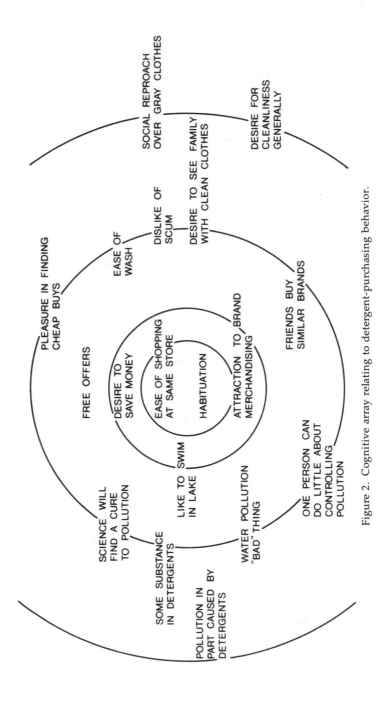

Figure 2. Cognitive array relating to detergent-purchasing behavior.

relationship, I found that this kind of cognitive array was indeed present. In a small town in Central British Columbia, Canada, I interviewed a randomly selected sample of 100 housewives to ascertain their views on water pollution and environmental deterioration generally, and their detergent purchasing behavior. During the period of the investigation (summer 1972) Canadian consumers were offered the choice between "nonpolluting" nonphosphate detergents and "polluting" phosphate brands. (Phosphates are powerful cleaning agents that also soften hard water, but provide one of many nutrient sources for primary biological plant life, especially algae. The resulting acceleration of primary biological production is known as eutrophication and is particularly prevalent in areas of shallow still water, especially lakes.) Such a choice had been available to the housewives investigated for six months prior to the survey, and the local newspaper had carried a number of articles about the possible dangers of eutrophication to the nearby lake into which the community discharged its untreated sewage.

The housewives were asked to respond to a number of attitude statements about water pollution, and then asked to explain why they bought the particular detergent they used. Table 1 lists the apparent "discrepancies" between overt behavior and attitude responses. It will be quickly spotted that the bulk of the apparent inconsistencies relate to socially conventional responses about environmental quality, but

TABLE 1

POSSIBLE "INCONSISTENCIES" BETWEEN VERBAL ATTITUDE AND BEHAVIOR AS FOUND IN THE DETERGENT STUDY

	N	%
Housewives purchasing phosphate detergents	61	100
(a) who stated that they were "concerned" or "very concerned" about water pollution as a problem	39	64
(b) who also believed that the individual should sacrifice a small amount to help clean up pollution	26	43
(c) who also stated that the nearby lake was of "low" or "very low" quality	14	23
(d) who also expressed a willingness to pay for an environmentally less damaging detergent	10	18
(e) who also knew that there was "some substance" in detergents that pollute water	5	9
(f) who also admitted that the detergent they were using polluted the water	3	5

that few of the women interviewed who bought phosphate-based detergents knew much about the relationship between detergents and eutrophication (though many admitted that they knew about some "substance" that might give rise to pollution). This will be elaborated below: of relevance to the discussion at this point are the data in Table 2, which show that phosphate-detergent users purchased their brands for a variety of nonenvironmental reasons. Under further probing, it became evident that, for the vast majority of phosphate-detergent users, habitual buying indicated satisfaction with the cleaning qualities of the product. The dilemma facing those housewives who knew (or cared to find out, or admitted to the interviewer) of the connection between phosphates and pollution of the nearby lake was that the nonphosphate product definitely did not clean as well, it left a nasty scum on the clothes and around the inside of the washing machine so that extra rinsing was necessary, and it was usually slightly more expensive than the phosphate brands (in terms of gross weight, but not necessarily in terms of quantity used per wash). Thus, it appears that of a number of cognitions associated with the complex phenomenon known as water pollution, only a few are related to the act of purchasing a detergent. But other, more powerful, cognitions are significant at such a time, hence the "discrepancy" (see Figure 2).

This kind of evidence highlights dangers inherent in personality testing devices like McKechnie's (1972) technique which confronts the respondent with some 330 attitude statements, all of which must be answered spontaneously along a "strongly agree–strongly disagree" scale. The results are factor-analyzed into major environmental dispo-

TABLE 2
REASONS WHY HOUSEWIVES CHOSE A PARTICULAR BRAND OF DETERGENT

	Phosphate detergent users		Nonphosphate detergent users	
	N	%	N	%
Effect on water pollution	—	—	29	74.5
Cleans well	25	41.0	1	2.5
Cheap	18	29.5	1	2.5
Cold-water use	6	9.8	—	—
Offers associated with brand	2	3.3	—	—
Other reasons	1	1.6	1	2.5
Always used the brand (i.e., no particular reason)	9	14.8	7	18.0

sitions. But the attitude battery is preselected and the evaluation depends upon self-description of environmental likes and dislikes. When it is realized that there is a behavioral equivalent to the cognitive array depicted in Figure 2, it will be seen that even the sophisticated response inventory cannot explain why people behave the way they do. Figure 3 illustrates this for the case of an individual confronted with the uncertain probability of suffering loss as a result of a natural hazard. A number of behavioral options to avoid or lessen the loss will be possible but only a few of them may be known, or available, and even fewer will be adopted, given the uncertainty of the event and the possible disruption caused by taking avoiding action. For each individual, some of the behavioral options will be known, some will be obscure, and some will be barred due to financial, social, and political factors beyond the subject's control. So there may not be an inconsistency between reported attitude and observed behavior, simply a failure or an inability to consider other possible behaviors.

Wicker's (1969, pp. 69–74) second set of latent process variables, namely, the environmental or situational factors, help to determine what alternative behaviors will be considered. Campbell (1963) believes that behavior will be influenced by the situational threshold,

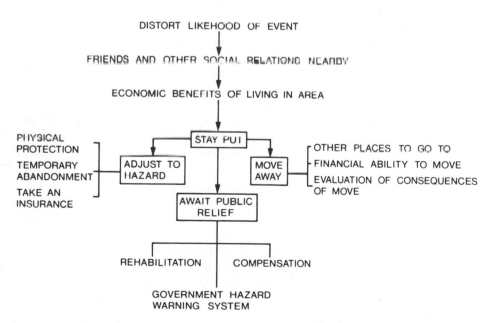

Figure 3. Behavioral options to cope with hazard.

beyond which there is a search for alternative behaviors. This idea is akin to the notion of behavioral thresholds currently being investigated in the natural hazards research program (Burton *et al.*, 1976), those critical points of perceptual discontinuity where the hazard-prone adopt a new kind of damage-avoiding behavior. We still don't know enough about these thresholds; certainly they are influenced by experience, the acquisition of new information, and exhortation from social peers. The presence of a peer group, especially if the subject knows that the peer group is aware of the subject's feelings on a given issue (e.g., littering) certainly influences the situational threshold, as do decisions of social policy that are explicitly aimed at altering public behavior (e.g., the addition of a refundable deposit on a beer can). Of special interest is the observation of behavior when a totally nonexperienced event occurs. How do people respond during periods of unanticipated environmental stress, such as a severe natural event or a dangerous pollution episode or an energy shortage? Returning to the energy case study introducing this chapter, Murray and his associates reported (1974, p. 260) that the novelty of the energy crisis lasted a mere week; after that frustration and anger increased dramatically. The new behavioral arena was intolerable for three quarters of those interviewed.

A third set of mediating variables that have both personal and situational components is what Fishbein (1967, p. 490) calls "social response influences" or "perceived social sanctions" that motivate the individual to comply with certain socially desirable codes of conduct. This is a crucial area of interest in environmental behavior research, for environmentalism is in essence the execution of social responsibility. Heberlein (1972, p. 81) postulates that socially acceptable behavior will be guided by three determinants. First, there are the attitudes toward the behavior itself, particularly the likely consequences of any action (or inaction) and the evaluation of these consequences. This can be termed the "knowledge component," which is a function of experience, education, and the cognitive ability to differentiate (i.e., to identify complex cognitive relationships) and to discriminate (i.e., the perceptual clarity with which differentiation is related to other behaviors).

Heberlein's second factor is the "culpability component," the attribution of blame for one's actions, and the identification of options that are less socially detrimental. Environmentally disruptive behavior is often tolerated because people feel they "have no choice" or because "everybody is doing it." Some social historians believe that the modern corporate and welfare state has taken away much of the sense of personal responsibility in work and play, leaving few if any social norms to guide conduct. Antisocial behavior (e.g., vandalizing recreational areas) is therefore not discouraged. In hazard-response studies Kirkby (1973) and Sims and Baumann (1974) found that behavioral adjustment to hazard was bound up with beliefs about fate. Those who felt that they had little or no control over their lives tended to accept the threat of hazard, and hence suffered more damage and loss of life, while those who believed that they could determine the course of their lives were more likely to adopt one or more behavioral adjustments. Culpability is evaluated by merging the pattern of cognitive differentiation and discrimination (Figure 2) with the mental schema of alternative possible behaviors (Figure 3). The skill in merging these two cognitive schema, known as integrative complexity, is obviously a tremendously important facility in environmental cognition and an important objective in environmental education.

In the detergent case study cited above, it was apparent that knowledge of consequence and culpability were major factors differentiating the phosphate detergent users from the nonphosphate product purchasers (Table 3). In an attempt to test for the effect of knowledge in influencing environmental behavior, two articles were prepared for the local newspaper. The first was a factual price identifying the known relationship between phosphates and eutrophication and pre-

TABLE 3
AWARENESS OF THE ROLE OF PHOSPHATES BY HOUSEWIVES

	Phosphate detergent users		Nonphosphate detergent users	
	N	%	N	%
No awareness	26	42.6	6	15.4
Minor awareness—knew that there was some "substance" in detergents contributing to water pollution	26	42.6	20	51.2
Moderate awareness—knew that phosphates did something "wrong" to water quality: mentioned algae	5	8.1	9	23.0
High awareness—knew of the whole process of eutrophication and its effects	4	6.6	3	7.7

$(X^2 = 9.751, P = 0.02, d.f. - 3)$

senting the phosphate concentration for all the locally sold brands. The second was an editorial stressing the importance of the information provided in the earlier article and outlining the environmental consequences of purchasing each of the two kinds of soap powder. In a follow-up survey of another 100 housewives a month later, it was evident that those already purchasing the nonphosphate products were more likely to have remembered and recalled the two articles (a copy was shown at the end of the interview) than were the phosphate detergent buyers at whom the information was aimed. Upon further discussion it was apparent that those for whom the whole question of water pollution and the effects of the detergent were nonsalient did not bother to read the articles at all, nor did they feel particularly guilty about their deed (Table 4).

Thus, it seems that environmentally responsible behavior tends to be associated with information search to justify and support the act. And this in turn leads to a more consistent and coherent cognitive array. This finding is in agreement with Festinger's (1957) dissonance theory in which attitudes are changed to be consonant with behavior. If this is true, it also helps to explain why information campaigns do not always succeed in changing either attitudes or behavior. After all, they are predicated on the dubious assumptions (1) that man is a rational and consistent creature who will modify his motivations on the basis of new information, say, from those of self-interest to altruism, (2) that by concentrating upon one set of cognitions other relevant cognitions will change, and (3) that attitude shift will be sufficiently profound and long-lasting to influence behavior.

Information seeking to reinforce environmentally responsible be-

TABLE 4
KNOWLEDGE AND RECOLLECTION OF NEWSPAPER ARTICLES BY HOUSEWIVES

	Phosphate detergent users		Nonphosphate detergent users	
	N	%	N	%
Did not remember or recollect articles	33	54.1	15	38.5
Recollected when shown articles	7	11.5	10	25.6
Remembered articles before being shown them	12	19.7	9	23.1

$(X^2 = 15.695, P = 0.05, d.f. = 2)$
N.B. Data do not add up to 100 as 16 respondents did not read the local paper.

havior will be influenced by the third constituent of the social response set, the "normative component." This centers on the relationship between what one personally feels is appropriate behavior, beliefs about what society (or at least significant others) feel about it, and beliefs about what ought to be done in the communal interest. This judgment will clearly be influenced by the social orientation of the individual, or, as Fishbein (1967, p. 488) puts it, his "motivation to comply with the norm, that is, his desire or lack of desire to do what he thinks he should do."

The forces that shape social obligation are most complicated and as yet ill-understood. Certainly the social milieu (friends, peer group influences, the opinions of respected individuals) is crucial, but so too are institutional factors, such as the degree to which work, school, or the home environment encourage a sense of responsibility and belonging. A variety of social critics (e.g., Roszak, 1973; Thompson, 1973; Rattray Taylor, 1974) have extended the notion of anomic (or lack of social norms) to all aspects of modern existence from work to leisure, classroom to the domestic hearth. Our modern technocratic–bureaucratic culture strips almost everybody of his or her sense of identity and personal worth, so that for many a sense of individual purpose and committed public-spiritedness is largely absent. Rattray Taylor (1974, p. 187) notes the implications for the modern acquisitive society:

> To no small extent our conception of ourself is impressed by the objects with which we surround ourself [sic]. These are not only evidences of achievement, they are status symbols in the proper sense: indications of our tastes and interests and our "walk of life." This is why it is so hard to get rid of the affluent society: stripped of his peculiar possessions a man becomes indistinguishable from other men; he loses his identity.

Both the energy and detergent case studies reveal that the majority only pay lip service to social pressures when they are aware that "most people they know" disregard such pressures. In this context it is interesting to speculate that those with a more public-spirited commitment may be more ready to exert the sanctions. In the detergent case study, for example, housewives were asked how they would react to a neighbor using a phosphate brand. Their replies, detailed in Table 5, indicate that those who already have switched to the less polluting varieties were more willing to comment on his or her behavior than were those who used the phosphate brands. But these may be "socially acceptable" responses; I have no idea of their actual behavior in such situations. In all probability many of the social peers of the nonphosphate detergent-using group probably had switched,

TABLE 5
RESPONSE BY HOUSEWIVES TO THE QUESTION, "IF YOU REALIZED THAT YOUR
NEIGHBOR WAS USING A PHOSPHATE DETERGENT, WHAT WOULD YOU DO?"

	Phosphate detergent users		Nonphosphate detergent users	
	N	%	N	%
Nothing	49	80.3	17	43.6
Tell her of the relationship	9	14.7	13	33.3
Exhort her to change detergents	2	3.3	8	20.5

$(X^2 = 15.695, P = 0.001, d.f. = 2)$

anyway. Nevertheless there is plenty of evidence that social norms are socially transmitted, so failure to "prick a neighbor's environmental conscience" could well contribute to the declining influence of such norms in personal behavior.

Eckhardt and Hendershot (1967) have produced a helpful model that might be used to test further the relationship between perceived social norms and environmentally sensitive behavior (Figure 4). Superficially the matrix is a simple variation of the twin themes of cognitive dissonance and cognitive consistency, the striving to compliment belief and action. In Box A the individual is acting in a manner that is consistent with his own cognitive set and consonant with his interpretation of social norms. This is nonstressful behavior

		PERSONAL INFLUENCES	
		BEHAVIOR CONSONANT WITH BELIEFS	BEHAVIOR DISSONANT WITH BELIEFS
SOCIAL INFLUENCES	BEHAVIOR CONGRUENT WITH PERCEIVED SOCIAL NORMS	A	B
	BEHAVIOR NON-CONGRUENT WITH PERCEIVED SOCIAL NORMS	C	D

Figure 4. Individual and social sanctions on environmental behavior. (After Eckhardt and Hendershot, 1967.)

that accounts for most of our culturally molded everyday acts. In Box B, the behavior is socially acceptable but personally undesirable; here would fit examples of socially obligatory environmental conduct such as unwillingly denying oneself a third child or sacrificing one meat meal per week. Because such an act is executed more for reasons of social conformity than personal preference, supportive public opinion and relevant information is actively canvassed. Here we return to the theme that information search and sensitivity to social norms may be triggered by dissonance-creating behavior, which subsequently swings environmentally consistent attitudes into line.

In Box C the conduct is socially unacceptable but personally desirable, such as having the third child or taking a shortcut through an ecologically fragile wilderness. In such circumstances that act is "rationalized" either by "switching off" perceived social sanctions, downplaying the social value of people voicing such sanctions, or distorting public opinion to suit the context of the act. This kind of situation is often evident in environmental policy issues when "Establishment" politicians are faced with opposition from environmentalist groups who purport to speak "for the people." For example, the politicians seek refuge in the advice of their principal consultants and "informed experts" whose credibility may be exaggerated beyond all proportion to their professional competence, while dismissing the social (and political) significance of the opposition. When confronted by an angry group of people (drawn from all classes of society), disturbed about a proposal for a waterfront freeway in Vancouver, the prodevelopment, profreeway mayor was quoted as describing the protesters as "Maoists, Communists, pinkos, left wingers, and hamburgers [defined as "persons without university degrees"]" (Guttstein, 1975, p. 165). In urging the Vancouver City Council to support the controversial freeway proposal, he claimed that should it flounder it would be a victory for the Communist party of Canada. In the end politics, not psychology, prevailed; the freeway was not constructed and the mayor, plus most of his prodevelopment colleagues, lost their seats in the next municipal election (Guttstein, 1975). A Box C situation is also evident when professional advisers are confronted by what they regard as "lay" opinion regarding technical matters such as nuclear reactor safety. Almost instinctively, they tend to discount such information and seek support from their professional peers; hence the reason why knowledgeable environmental activist groups hire or otherwise obtain the views of professionally competent people when preparing their cases.

In many respects there are uncomfortable similarities between

Boxes A and D even though they appear to be diametrically opposed. Admittedly, at one extreme the individual would be schizophrenic, totally divorcing conduct from beliefs, since his behavior is both inconsistent with his beliefs and incompatible with his perception of social norms. But if the cognitive elements are peripheral, Box D represents most of our everyday "antienvironmental" behavior: driving cars and not maintaining pollution-control devices, trampling down bushes when the proper trail should be followed, consuming environmentally damaging goods when more expensive and time- and effort-consuming environmentally "sound" substitutes are available, or urinating in a bathing area even when there is a public toilet nearby. As already discussed, this kind of behavior is common because social sanctions may not be present, or the knowledge and culpability components are weakly developed. Thus, millions of people in millions of ways add incrementally to the deterioration of the collective environmental amenity while professing a sense of environmental concern.

THE BEHAVIOR-COGNITION LINK

In the light of the very sketchy evidence that attitudes do influence behavior, a number of psychologists have postulated that perhaps the directional pulse is the wrong way around and that by changing a person's behavior he or she will draw a new set of inferences about what he or she feels and believes. Bem (1971, pp. 54–69) has introduced and tested the self-perception theory, which states that attitudes follow from behavior and can be inferred from observing behavior. Heberlein (1972, 1976) goes even further. He believes that people should be guided into new behavior patterns through policies that alter present environmentally damaging conduct or change the social conditions in which it occurs. He cites the example of barring certain kinds of boating from some lakes, then providing explanatory information. If this information can be phrased in such a way as to be consonant with other motivations, the result should be quite felicitous. He also advocates (1976) more extensive use of the quasi-experiment as an environmental policy mechanism, the essence of which is to simulate behavioral change and test for the effect on attitudinal thresholds. For example, selected samples of people could be told that a surcharge (a different amount for each group) was going to be placed upon their tax bill to provide sufficient funds to control water pollution in their area. Each group would then be offered a variety of behavioral options (public meetings, letters of protest, petitions, etc.) with which to respond. Depending on what alternatives they chose and how they made use of them, policymakers would then have a clearer idea of public willingness to pay for pollution control. Naturally, certain legal and ethical issues would have to be properly considered first, but the idea has merit in that it links behavioral modification to information-seeking.

Bruvold (1973, pp. 214–215) makes further useful suggestions in this regard. Because only weak relationships exist between any *one* attitude and a particular behavior, and any *one* behavior and a particular attitude, a policy to encourage environmentally compatible behavior should aim at the relevant clusters of cognitions and behaviors (some positive and some negative) to make them all favorable to appropriate behavior change. As an illustration he suggests that pecuniary incentives (by means of tax relief) should be made available to encourage householders to overcome their reluctance in sorting their solid waste into easily recyclable or disposable categories. This, in combination with information about the hazards and costs of not so doing, plus facts about the environmental and economic gains of proper solid waste management might lead to the birth of a new waste removal program.

Certainly there is evidence that combining the motivations can lead to environmentally sound behavior, though how far this leads to environmentally consistent attitudes, especially in other behavioral arenas, remains in doubt. In the detergent case study, for example, it was found that all but a few housewives recycled their soft drink cans and bottles, regardless of their detergent-using behavior. The results in Table 6 reveal, however, that most of the phosphate-detergent users

TABLE 6
REASONS FOR DISPOSING OF SOFT DRINK CANS AND BOTTLES BY HOUSEWIVES

	Phosphate detergent users		Nonphosphate detergent users	
	N	%	N	%
Put in garbage	4	6.6	1	2.5
Money for children	44	71.8	20	50.8
Specifically for recycling	9	15.0	16	41.2
No answer	4	6.6	2	5.5

(X^2 not significant)

did so to give money either to their own children or to children collecting such containers. A much greater percentage of the nonphosphate detergent users claimed they recycled for environmental reasons.

Clearly there is a tremendous need for more field research into the behavior–cognition link. Certainly there will be "behavioral thresholds" beyond which people will not willingly pass and their cognitive orientations will not shift. People did not join Temperance societies following Prohibition, nor are they vehemently supporting clean air policies now that they have to pay for emission control devices. Wall (1973) did not find that those who were forced to switch to smokeless heating fuels under the British Clean Air Act (1956) were noticeably more concerned about air pollution than those who were still burning coal. In any case, we cannot assume that by concentrating on one aspect of observed behavior, connected behaviors will necessarily be consistent. For example, because people join environmental action groups, there is no reason to expect that their actions will be uniformly proenvironmental. Barnett (1971) found that members of Zero Population Growth, Inc. were unable to reconcile their personal interest in family life with their profound belief that America is overpopulated and hence that some couples should have no children at all. Hendee (1972) commented that some of the worst ecological damage to the western American wilderness areas was done by Sierra Club outings that consisted of far too many people for such ecologically sensitive terrain. (Both the Sierra Club and the U.S. Forest Service, which is responsible for such areas, now restrict group size.)

The behavior → attitude model is also unsatisfactory in that it assumes few, if any, impediments to behavioral choice. But the fact remains that many people who suffer environmental deprivation due to social, economic, or political factors beyond their control may well be forced to adopt consistent cognitive arrays against their better judgment (Svart, 1974). For example, McLean (1974) found that mothers living in a low-quality urban environment were deeply unhappy about the sordid physical environment that surrounded them, but had to accept it as their fate. Accordingly, they valued highly the proximity of relatives and friendly neighbors, whose companionship provided a crucial sense of social belonging. Middle-class suburbanites, on the other hand, who have moved to their neighborhoods by their own choosing and who enjoy social connections throughout the metropolitan community, worry far more about the appearance of the surrounding property and any possible factors that might lower property values.

Many people who live in hazard-prone locations may be there not because they do not know of the dangers of the hazard, but because hazardous places are cheap to buy or to squat in. Those who suffer toxic chemical pollution from motorways or factories may be imprisoned into ill health because no alternative housing is available. Yet, if asked about it, all these people may play down the danger to their well-being, because it is psychologically and socially uncomfortable to do otherwise. It is therefore dangerous to isolate environmental cognition from the wider sociopolitical milieu in which people live.

THE TRANSACTIONAL MODEL

We still know very little about the interactions between information flows, experience, awareness, and concern for the well-being of others in shaping cognitions and influencing behavior. But it now seems unlikely that either attitude → behavior or the behavior → attitude model is entirely suitable, for neither properly takes into account the social and political forces that play such a major role in molding cognitions and restricting behavioral options. A more plausible model, then, is some kind of transactional arrangement in which the individual "negotiates" with environmental stimuli in a symbiotic manner, each influencing the other. Environmental behavior thus becomes something of a "game" in which the individual is constantly testing both the environmental response and the reaction of the social-political system in which he or she operates. Simultaneously, by this process, the subject learns more about the limits of his or her own abilities and the barriers created by social and political forces beyond his or her control. The transactional model of environmental cognition and behavior, therefore, describes a process of individual "becoming" and the recognition of political consciousness, while demonstrating that there are limits to man's alteration of his surroundings.

Although there have been no field studies to test this model, Kunreuther (1974) has begun to work on a version of it in his natural hazard research studies. From Figure 5 it will be seen that when trying to avoid damage or injury caused by natural hazard, most individuals pass through a series of explorations of options. The range of these explorations and the assessment of the choices will depend upon such factors as the characteristics of the hazard (predictability, warning time, force, duration, etc.), previous experience by the individual of the hazard or other hazards, knowledge of existing public policy, and the practicability of avoidance. Whatever option is chosen will clearly

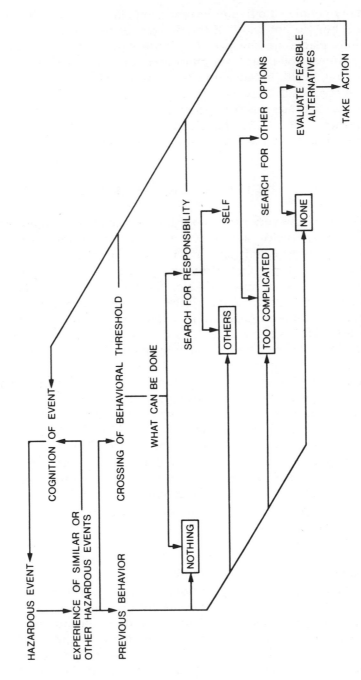

Figure 5. Ordered choice under uncertainty. (Based on Kunreuther, 1974.)

alter a number or cognitions about the hazard: The hazard, then, is not a probabilistic event with defined physical characteristics, but a subjectively appraised phenomenon whose meaning alters with the context of choice.

The transactional model of environmental cognition and behavior holds much promise since it recognizes that in cognizing and responding, man learns to cope with his world, and to change things a little to suit his requirements. Conceptually, the model bridges the mechanistic and rationalistic attitude behavior schema depicted in Figure 1a and the interactional arrangements of the suggestions portrayed in Figures 1b and 1c where feedback is postulated but is narrowly defined. Thus, cognitive structuring and response are placed in the real-world framework of possibilities and impediments, hopes and fears, privilege and inequality.

In its essence, the transactional model is one of environmental consciousness, and environmental consciousness cannot be separated from political efficacy. There is now overwhelming evidence in the literature that expressions of environmental concern and the actions of radical environmental reformers are linked to a strong sense of political competence, (see, for example, Constantini and Hanf, 1972; Dillman and Christenson, 1972; Tognacci et al., 1972). It is not just the obvious fact that as a person's private environment becomes more secure (high social mobility, an enjoyable well-paid job, a nice house, and good education for one's children), he or she can afford to be bothered by "cosmetic" disamenities such as pollution, noise, and visual blight. It is that such people move in circles of political influence: they firmly believe that if necessary they are capable of changing the forces that affect their surroundings. The poor and other exploited people, on the other hand, cannot afford the luxury of worrying about environmental disruption, for their priorities rest with the necessities of life—food, jobs, housing, and education—and in any case they face higher political costs when seeking to improve their environmental quality. The lack of a decent life is *the* most important personal and social issue for such people, most of whom feel quite powerless to change their lot. In this context it is interesting to speculate which of the two groups of modern postindustrial society, the environmentally privileged and the environmentally exploited, are more likely to display inconsistency between their environmental words and deeds. Most probably the former group, because surely the real issue in modern environmentalism is why the politically powerful prefer to worry more about the eradication of pollution and other manifestations of disamenity (at negligible sacrifice to their mode of existence but with great potential gains to their amenity or property values) rather than confront the social injustices that are central to the environmental dilemma. Participatory activists such as Friere (1970), Starrs and Stewart (1971), and Kasperson and Breitbart (1974, pp. 41–59) argue that this crucial issue can only be resolved by means of a systematic program of politicizing the public to be aware of their environmental rights and the political influences that ensure exploitation. It may thus be possible to postulate a program of research and education that links the conceptual value of the transactional model to a policy of political consciousness-raising. Recent advances in activist environmental education and the widening of public input through the statutory medium of environmental impact assessment should help to open up this possibility.

IMPLICATIONS FOR PUBLIC POLICY

Because the relationship between environmental cognition and behavior is linked to wider political considerations, it will be useful to speculate on some of the implications of this relationship for future environmental policymaking.

METHODOLOGY

The traditional mechanism used to gauge public feeling on a matter before enacting policy is the opinion poll. In view of the information provided in this chapter, it seems desirable that questionnaire techniques should be reassessed very carefully indeed. Most polls only monitor weakly held cognitions, mostly of a socially acceptable kind, regarding matters that most respondents have thought little about (especially the various implications of their statements), but to which they are required to give spontaneous answers, often a very limited array of rather simplistic choices. Erickson and Luttberg (1973, p. 17) assert that most Americans (about 83%) take very little interest in "social questions" such as pollution, housing, social and economic inequality, and demonstrate very little knowledge of the political alternatives being discussed. Converse (1964) found that shifts in public opinion over nonpersonal political issues can be explained as much by random factors and by the research design (including construction of the questionnaire and the nature of the interview) as by a genuine swing in public interest. He concluded that less than

20% of respondents held opinions that were meaningful to them, even though two thirds, when pressed, offered a viewpoint.

These findings help to explain the fickleness of public opinion and the relative ease with which it is swayed by media information. Munton and Brady (1970), Sprout (1971), and McEvoy (1973) have all shown a relationship between response to environmental policy issues and media coverage, though students of communications are quick to point out that most of this kind of information is selectively perceived, tending to reinforce existing prejudices. There is little doubt, for instance, that media reports of substantially increased oil company profits focused preexisting public antipathy toward the corporate establishment, but the result was sufficiently effective to produce quite noticeable political consequences, as demonstrated by the current impasse between Capitol Hill and the White House.

An additional problem with many so-called "perception" studies is the difficulty of obtaining a sensible response to hypothetical situations. Because few, if any, people can conceptualize all the possible outcomes of an anticipated course of action, inbuilt cognitive inconsistencies may become even more apparent. Winham (1972), for example, reported that the residents of Hamilton, Ontario, desired a reduction in air pollution, but not at the expense of economic growth or income protection. Gallup (1973) found that while 80% of Americans were aware of the energy crisis, and a majority favored stiffer taxes on oil depletion and tougher controls on speed, 66% regarded a "sizable piece of land" (up to one acre) as a very important feature in the selection of a home. But single-family houses are enormously energy demanding (unless radically redesigned), and all but impossible to service with low-energy transportation systems. Of course, the questionnaire helps to elucidate inconsistencies that already exist, but it does not always help policy analysis.

FUTURE RESEARCH

The research possibilities opened up by the transactional model are immensely challenging. Certainly it would be helpful to focus more attention on the motivations behind the behavioral side of the relationship, that is, less on what people do and more on why they do it. "When one knows why a particular behavior is preferred," notes Schiff (1971, p. 16), "one can attempt to change the behavior by changing the advantages seen accruing to it." This notion of "combining the motivations" deserves further exploration in the light of evidence that an information search leading to consistent cognitions is associated with behavioral change. The use of the quasi-experiment holds much promise so long as it is thoroughly screened against abuse. Where policy decisions affect behavioral options (as, for example, the use of permits and reservations in wilderness areas), cognitions and subsequent behavioral responses should be carefully monitored.

Public Participation

If it is accepted that a transactional or experimental view of environmental construing leads to improved self-awareness and political consciousness, then programs to involve key citizens in policymaking make sense. Ever since the pioneering study by Katz and Lazarfield (1965) that many opinions are socially transmitted by "key influentials," many social scientists now agree that society divides into a number of groups and coalitions that produce "leaders." If these people can be identified, then the workshops or task force becomes a viable mechanism for public participation. The workshop/task force device thus makes use of two fundamental characteristics of policymaking, namely, that political communication takes place through social networks, and that social problems based on community conflict are best solved by what Thayer (1971, p. 4) calls the "collegial, nonhierarchical, face to face, problem solving group, large enough to include the perspectives and expertise necessary to deal with the problem at hand, but small enough to ensure each participant that his or her contribution is substantial, meaningful and indispensable to the process." The task force is becoming more of an integral part of resource management, but it remains to be seen whether the current holders of political power will be sufficiently responsive to give it adequate rein (O'Riordan and Sewell, 1976).

Environmental Education

The essence of environmentalism is the resolution of opposing forces that urge us to execute actions that may conflict with our beliefs. Environmental education should be aimed in part at helping all generations confront this dilemma and hence become more aware of the true values of existence. It is uncomfortable but apparently necessary to recognize that one's words and deeds, or actions and moral beliefs are not always consistent. A student of mine once saw three youths, each of whom had dropped litter. She picked up their trash, followed them for a while, and then asked them their opinion of people who littered. The youths castigated such people as socially irresponsible and were dumbfounded when she produced the evidence of their own behavior. Two of the boys denied their actions completely, but the third was most contrite. Will this lesson influence their future littering behavior? I am convinced that the answer will depend on how each of the boys feels about his personal obligation to the society in which he lives and which in part shapes his view on the world.

From different vantage points, three authors contributing to the same symposium (Burger, 1974; Du Boff, 1974; and Morrison, 1974) all agree that the necessary social change toward a better mode of existence will only come through the patient dedicated work of "radical ecological activists" who proselytize within the system, who expose its inequities, and who offer constructive alternative proposals that, step by step, lead to a new pattern of political and social relationships. These people must help their neighbors "open their eyes" and organize effective political cooperatives that will obtain improvements in their well-being by peaceful and democratic means. An essential step in this process is to develop a force of social conscience that mediates the conflicting cognitive processes described in Figure 1. Back in 1949 Aldo Leopold called for a land ethic, a new morality toward nature. Today it is evident that this morality must encompass how we behave toward our fellow human beings, alive and yet unborn, if we wish to survive in a just, peaceful, and habitable world.

REFERENCES

Barnett, L. D. Zero Population Growth, Inc. *BioScience*, 1971, *21*, 759–265.

Bem, D. *Beliefs, attitudes and human affairs*. Belmont, Calif.: Brooks Cole, 1971.

Black, J. N. *The dominion of man: The search for ecological responsibility*. Edinburgh: John Black, 1970.

Bruvold, W. H. Belief and behavior as determinants of environmental attitudes. *Environment and Behavior*, 1975, *5*, 202–218.

Burch, W. R. Jr. *Daydreams and nightmares: A sociological essay on the human environment*. New York: Harper & Row, 1972.

Burger, W. Ecological viability: political options and obstacles. In H. G. T. Van Raay and A. E. Lugo (Eds.), *Man and environment ltd*. Rotterdam: Rotterdam University Press, 1974, 235–256.

Burton, I., Kates, R. W., and Kirkby, A. V. T. The cognitive reformation: geographical contributions to man–environment theory. *Natural Resources Journal*, 1976, in press.

Campbell, D. T. Social attitudes and other acquired behavioral dispositions. In S. Koch (Ed.), *Psychology: A study of a science*. Vol. 6. New York: McGraw-Hill, 1963, 94-172.

Constantini, E., and Hanf, K. Environmental concern and Lake Tahoe: A study of elite perceptions, backgrounds, and attitudes. *Environment and Behavior*, 1972, *4*, 209–242.

Converse, P. E. The nature of belief systems in mass publics. In D. Apter (Ed.), *Ideology and discontent*. Glencoe, Ill.: Free Press, 1964.

Crespi, I. What kinds of attitude measures are predictive of behavior? *Public Opinion Quarterly*, 1972, *35*, 327–334.

De Fleur, M. L., and Westie, F. R. Attitude as a scientific concept. *Social Forces*, 1963, *42*, 17–31.

Deutscher, I. Words and deeds: Social science and social policy. *Social Problems*, 1966, *13*, 235–265.

Dillman, D. A., and Christenson, J. A. The public value for pollution control. In W. R. Burch, Jr., N. H. Cheek, and L. Taylor (Eds.). *Social behavior, natural resources and the environment*. New York: Harper & Row, 1972, 237–256.

Drew, E. B. Dam outrage: The story of the Army Engineers. *Atlantic Monthly*, 1970, April, 51–62.

Du Boff, R. B. Economic ideology and the environment. In H. G. T. Van Raay and A. E. Lugo (Eds.). *Man and environment ltd*. Rotterdam: Rotterdam University Press, 1974, 201–220.

Eckhardt, K. W., and Hendershot, G. Dissonance, congruence and the perception of public opinion. *American Journal of Sociology*, 1967, *73*, 226–234.

Erickson, R. S., and Luttberg, N. A. *American public opinion: Its origins, content and impact*. New York: Wiley, 1973.

Festinger, L. A. *A theory of cognitive dissonance*. Stanford: Stanford University Press. 1957.

Fishbein, M. Attitude and the prediction of behavior. In M. Fishbein (Ed.), *Readings in attitude theory and measurement*. New York: Wiley, 1967, 467–492.

Ford Energy Policy Project. *A time to choose*. New York: Ballinger, 1974.

Friere, P. *The pedagogy of the oppressed*. New York: Herder and Herder, 1970.

Galbraith, I. K. *Economics and the public purpose*. Boston: Houghton Mifflin, 1973.

Gallup, G., Jr. What do Americans think about limiting growth? Princeton: Public Opinion Research Center, 1973.

Glacken, C. L. *Traces on the Rhodian shore: Nature and culture in Western thought*. Berkeley: University of California Press, 1967.

Guthrie, D. A. Primitive man's relationship to nature. *Bio Science*, 1971, *21*, 721–723.

Guttstein, D. *Vancouver Ltd*. Toronto: Lorimer, 1975.

Heberlein, T. A. The land ethic realized: Some social psychological explanations for changing environmental attitudes. *Journal of Social Issues*, 1972, *28*, 79–87.

Heberlein, T. A. The three fixes: technological, cognitive and structural. In J. Field, (Ed.). *Water and community development: Social and economic perspectives*. 1974.

Heberlein, T. A. Some observations on alternative mechanisms for public involvement: The hearing, the public opinion poll, and the quasi experiment. *Natural Resources Journal*, 1976, *16*, 197–212.

Hendee, J. C. Personal communication, 1972.

Kasperson, R. E., and Breitbart, M. *Participation, decentralization and advocacy planning*. Washington: Association of American Geographers, 1974.

Katz, E., and Lazarfield, P. E. *Personal influence: The part played by people in the flow of mass communication*. New York: Free Press, 1965.

Keisler, C. A., Collins, B. E., and Miller, N. *Attitude change: A critical analysis of theoretical approaches*. New York, Wiley, 1969.

Kirkby, A. V. T. Some perspectives on environmental hazard research. Discussion Paper University of Bristol: Department of Psychology, 1973.

Kunreuther, H. Protection against natural hazards: A lexicographic approach. Discussion Paper 45, Fels Center of Government, University of Pennsylvania, 1974.

Landsberg, H. H. Low cost abundant energy: Paradise lost? Resources for the Future Annual Report, 1973, 27–52.

Lauer, R. H. The problems and values of attitude research. *Sociological Quarterly*, 1971, *12*, 347–352.

Leiss, W. *The dominion of nature*. New York: Brazillier, 1972.

Leopold, A. *A sand country almanac*. New York: Oxford University Press, 1949, 201–226.

Martin, P. S. Prehistoric overkill. In P. S. Martin and H. E. Wright, Jr. (Eds.), *Pleistocene extinctions: The search for a cause*. New Haven: Yale University Press, 1967.

McLean, U. Environmental perceptions in a deprived area. In J. T. Coppock and C. B. Wilson (Eds.), *Environmental quality*. Edinburgh, Scottish Academic Press, 1974, 178–188.

McEvoy, J. III. The American public's concern with the environment. In C. R. Goldman, J. McEvoy III, and P. J. Richerson, (Eds.), *Environmental quality and water development*. San Francisco: Freeman, 1973, 135–156.

McKechnie, G The environmental response inventory. Berkeley: University of California, Institute of Personality Assessment, 1972.

McLuhan, T. C. *Touch the earth: A self-portrait of Indian existence*. New York: Pocket Books, 1972.

Morrison, J. F. Man, organization and the environment. In H. G. T. Van Raay and A. E. Lugo, (Eds.), *Man and environment ltd*. Rotterdam: Rotterdam University Press, 1974, 177–200.

Munton, D. & Brady, L. American Public Opinion and Environmental Pollution. Behavioral Science Laboratory, Ohio State University, 1970.

Murray, J. R., Minor, M. J., Bradburn, N. M., Cotterman, R. F., Frankel, M., and Pisarski, A. E., Evolution of public response to the energy crisis. *Science*, 1974, *184*, 257–264.

Nordhaus, W. D. 1974. Resources as a constraint on growth. *American Economic Review*, 1974, *64*, 22–26.

Otaiba, H. E. M. *OPEC and the petroleum industry*. London: Croom Helm, 1975.

O'Riordan, T., and Sewell, W. R. D. The culture of participation in environmental decision making. *Natural Resources Journal*, 1976, *16*, 1–22.

Pirages, D. C., and Ehrlich, P. R. *Ark II: Social response to environmental imperatives*. San Francisco: Freeman, 1974

Rappoport, R. A. *Pigs for the ancestors*. New Haven: Yale University Press, 1967.

Rattray Taylor, G. *Rethink: Radical proposals to save a disintegrating world*. London: Penguin, 1974.

Ridker, R. G. (Ed.) *Population, resources and environment*. Washington: Government Printing Office, 1973.

Rokeach, M. *Beliefs, attitudes and values*. San Francisco: Jossey Bass, 1968.

Roszak, T. *Where the wasteland ends: Politics and transcendence in a postindustrial society*. New York: Anchor, 1973.

Schiff, M. Some considerations about attitude studies in resource management. Department of Geography, Waterloo Lutheran University, Ontario, 1971.

Sims, J. H., and Baumann, D. The tornado threat: coping styles of the north and south. In J. H. Sims and D. Baumann (Eds.), *Human behavior and the environment*. Chicago: Maaroufa Press, 1974, 108–125.

Sprout, H. The environmental crisis in the context of American politics. In L. L. Roos, Jr. (Ed.), *The politics of ecosuicide*. New York: Holt, Rinehart & Winston, 1971, 41–50.

Starrs, C., and Stewart, G. Gone Today and Here Tomorrow: Issues Surrounding the Future of Citizen Involvement. Toronto: Ontario Government Queen's Printer, 1971.

Svart, L. On the priority of behaviour in behavioural research: A dissenting view. *Area*, 1974, *6*, 301–5.

Tarter, D. E. Looking forward: The case for hard-nosed methodology. *American Journal of Sociology*, 1970, *5*, 276–278.

Thayer, F. C. Participation and liberal democratic government. Toronto: Ontario Government Queen's Printer, 1971.

Thomas, K. (Ed.) *Attitudes and behaviour*. London: Penguin, 1971.

Thompson, W. I. *Passages around the earth: An exploration of a new planetary culture*. New York: Harper & Row, 1973.

Tognacci, L. N., Weigel, R. H., Wideen, M. F., and Vernon, D. A. T., Environmental quality: How universal is public concern. *Environment and Behavior*, 1972, *4*, 73–86.

Tuan, Y. F. Discrepancies between environmental attitudes and behaviour: Examples from Europe and China. *Canadian Geographer*, 1968, *12*, 176–181.

Tuan, Y. F. Our treatment of environment in ideal and actuality. *American Scientist*, 1970, *50*, 244–249.

Tuan, Y. F. *Man and Nature*. Washington: Association of American Geographers, 1971.

Tuan, Y. F. Structuralism, existentialism and environmental perception. *Environment and Behavior*, 1972, *6*, 319–331.

Tuan, Y. F. *Topophilia: A study of environmental perception, attitudes and values*. Englewood Cliffs: Prentice-Hall, 1974.

Van Arkel, D. Society and technology: 30,000 years in shorthand. In H. G. T. van Raay and A. E. Lugo (Eds.), *Man and environment ltd*. Rotterdam: Rotterdam University Press, 1974, 19–32.

Wall, G. Public response to air pollution in south Yorkshire, England. *Environment and Behavior*, 1973, *5*, 219–248.

Weisberg, B. *Beyond repair: The ecology of capitalism*. Boston: Beacon Press, 1971.

White, L., Jr. The historical roots of our ecologic crisis. *Science*, 1967, *155*, 1203–1207.

Wicker, A. W. Attitudes versus actions: The relationship of verbal and overt behavioral responses to attitude objects. *Journal of Social Issues*, 1969, *24*, 41–78.

Winham, G. Attitudes on pollution and growth in Hamilton, or 'there's an awful lot of talk these days about ecology.' *Canadian Journal of Political Science*, 1972, *5*, 389–401.

Winn, D. J. The psychology of smog. *The Nation*, March 5, 1973, 294–298.

2

Environmental Aesthetics: The Environment as a Source of Affect

JOACHIM F. WOHLWILL

For we all know that ugliness breeds ugliness, crime, corruption, disregard for law and order, disrespect for God and man, in short delinquency in all ages. The converse of ugliness—beauty—begets beauty, in all its manifestations, in nature, in man's handiwork, and in the realm of the spiritual.

The value of beauty and the price of ugliness can be reckoned in dollars. But the ways in which the enhancement of beauty and the abatement of ugliness—in city, village and countryside—can add to the inner prosperity of the human spirit are beyond calculation. (Governor's Conference on Natural Beauty, 1966, p. 7).

One might be tempted to dismiss the preceding statement, made at a Pennsylvania Governor's Conference on Natural Beauty, by the Conference Chairman, Frank Masland, Jr., as the sort of rhetoric expected of a keynote speaker at such a gathering. Yet, though there may be no scientific evidence whatsoever to support the rather sweeping assertion concerning the power of beauty and ugliness to exert such profound effects on human beings (and this writer knows of none), it may well reflect commonly held beliefs about the impact of the aesthetic quality of the environment on the individual. Such convictions have, of course, been expressed for centuries by philosophers, naturalists, writers, etc., and are certainly not limited to laymen, or

JOACHIM F. WOHLWILL · Division of Man–Environment Relations, The Pennsylvania State University, University Park, Pennsylvania.

politicians, even today, as reflected in the following quote from an environmentally concerned biologist in a letter in *Science*:

> . . . has there been, or will there soon be sufficient selection by polluted metropolitan environments to erase man's unspoken needs for open spaces, wild mountains, clean lakes, or small towns? Does Dobzhansky mean it is desirable to permit (let alone encourage) adaptation to New York-type cities, their bleak lifeless canyons of stone crawling with humanity, their noisy streets and overcrowded subways? . . . I don't know whether Dobzhansky has forgotten what it was like to walk the dunes in solitude or to swim in the ocean, but to most humans . . . it is pleasanter than basking in 5 P.M. traffic on Fifth Avenue (Iltis, 1967).*

While this statement is testimony to the intense feelings that our physical surroundings may evoke in us, the historical record of people's affective and evaluative reactions to their environment, both natural and man-made, does not by any means reveal the kind of consensus as to what is beautiful or ugly in our surroundings, and what effects different kinds of surroundings have on us that one would presuppose before taking statements such as the above at face value. Consider Samuel Johnson's response to the mountains:

> The Pyrenees were "not so high and hideous as the Alps," but "uncouth, huge, monstrous excrescenses of Nature, bearing nothing but craggy stones . . . An eye accustomed to flowery pastures and waving harvest is astonished and repelled by this wide extent of hopeless sterility" (quoted in Shephard, 1967, p. 131).

Similarly, indictments of the urban environment on the most diverse grounds including the aesthetic abound, as shown in the writings of many of the foremost literary figures in the United States of the 19th and early 20th centuries (cf. White and White, 1962); yet it is not difficult to call the roll of those who have risen to its defense, again basing themselves in part on aesthetic considerations (e.g., Cullen, 1961; Jacobs, 1961; Lynch, 1960).

In this respect, however, environmental aesthetics is clearly in the same position as the field of traditional aesthetics, i.e., of the response to works of art. The well-known variations among different cultural groups, as well as across periods of time, in the evaluation and affective response to paintings, music, dance, and the like, have not

---

* It may seem that Iltis, in these comments, is concerned with more than a purely aesthetic response to the environment on the part of the individual; yet, the complete text of the letter, while revealing more fully the perspective of the biologist interested in processes of evolution and adaptation, brings out clearly the importance that this writer accords to the stimulus environment, and its effects on the individual's well-being and satisfaction (cf. also Iltis, 1968).

kept psychologists and others from attempting to deal with this realm in a broader sense, that is, in a search for general principles, or—to borrow a term from the domain of linguistics—for universals. The present paper is predicated on the assumption that such general principles similarly exist in our affective responses to the physical environment, and that they may be closely related to those that apply in the domain of art, even though these may need to be modified and extended in important respects. The starting point for the treatment to follow will therefore be a general theory of psychological aesthetics that has centered on the power of particular properties of aesthetic stimuli to elicit affect. This theory will provide the framework for a review of the empirical literature in the area of environmental aesthetics, culminating in a critical examination of the theory and its application to this area. The remainder of the paper is devoted to a discussion of issues of methodology arising in research on environmental aesthetics, and an examination of some of the major unresolved questions in this field. The paper concludes with a brief section on applications to applied problems in the field of environment and behavior.

A PSYCHOLOGICAL FOUNDATION FOR ENVIRONMENTAL AESTHETICS: THE WORK OF BERLYNE

The single most influential contributor to the development of psychological aesthetics as an empirically as well as theoretically grounded field of investigation is undoubtedly Daniel Berlyne, who, along with a group of collaborators, has over the past decade been working on the systematic extension of his earlier seminal work, *Conflict, Arousal and Curiosity* (Berlyne, 1960) to the domain of aesthetics (Berlyne, 1972a; 1974). No comprehensive account of this impressive body of theory and research can be given here; suffice it to refer to two concepts of central importance, both of which appear of considerable relevance to the realm of environmental aesthetics. These are the concepts of "collative" properties of stimuli, and the distinction between "specific" and "diversive" exploration.

THE ROLE OF COLLATIVE PROPERTIES OF STIMULI

In Berlyne's (1960) original statement of his theory of the impact of stimuli on arousal and exploratory activity, a set of interrelated attributes of stimuli were postulated under the general rubric of collative properties, i.e., those that elicit either implicit or explicit comparative responses. Among these Berlyne listed complexity, novelty, incongruity, and surprisingness. For our purposes, a somewhat revised list will prove useful, which differentiates complexity into two related aspects, i.e., diversity and structural or organizational complexity, and adds an additional attribute, that of ambiguity, to Berlyne's set. The essential aspect of these several stimulus attributes is that all relate to the uncertainty contained within a stimulus, or the conflict it engenders in the individual in attempting to interpret it. Accordingly, a stimulus elicits investigatory or exploratory responses designed to reduce the uncertainty or conflict engendered by it, to the extent that it possesses such collative properties.

The conflict notion may be illustrated most directly with reference to the attributes of ambiguity, which is in effect defined as a conflict between competing alternative meanings or interpretations of a stimulus, and that of incongruity, which entails a mismatch between different aspects or portions of a stimulus, or between a stimulus and its context. Conflict may similarly be linked to surprisingness, which is basically the temporal equivalent of incongruity, based on a mismatch between some expected event and one actually encountered by the individual. In the case of complexity, the parent concept of uncertainty, in the informational sense, is more pertinent; indeed, for artificially constructed stimuli it is frequently possible to define complexity operationally in terms of some measure of structural information, i.e., uncertainty. By extension, uncertainty may generate conflict; i.e., the greater the informational content of a stimulus, the greater the conflict among alternative ways of identifying, classifying, or organizing it. Finally, in the case of novelty, the conflict notion remains largely implicit, although not entirely irrelevant, if we conceive of novelty as involving a discrepancy between the range of stimuli previously encountered by the individual and the particular characteristics of the stimulus presently confronting him.

It would not be difficult to demonstrate the direct relevance of each of the attributes just cited to the individual's response to the environment; the tourism industry itself provides ample testimony of the attraction that locales characterized by such attributes as diversity, novelty, and incongruity hold for many people. More specific consideration of the actual role played by these attributes will be deferred to the following section on empirical evidence relating to Berlyne's theory. At this point, however, it is necessary to face up to a question that is bound to arise in relation to the relevance of these collative properties in the context of a paper on environmental aesthetics. What do these attributes have to do with aesthetic quality, i.e., with beauty?

Psychologists, including Berlyne himself, are still some distance from being able to offer an explanation of the experience of beauty, which has been the subject of philosophical treatises since the days of Plato (e.g., Santayana, 1896). Indeed Berlyne did not introduce the collative properties just referred to in relation to this problem, but rather as determinants of spontaneous exploratory activity on the part of the individual, whether rat, monkey, infant, child, or college sophomore. The general hypothesis, verified through abundant research in Berlyne's laboratory and elsewhere, was that the tendency of the individual to engage in voluntary active exploration of a stimulus is proportional to the amount of uncertainty or conflict it engenders; it is thus highest for stimuli relatively rich in diversity, structural complexity, novelty, incongruity, or surprisingness. But this leaves us still somewhat removed from the problem of aesthetic quality; in order to bridge this gap, it will be helpful to introduce the second of Berlyne's concepts, the differentiation between specific and diversive exploration.

Two Exploratory Attitudes: Specific versus Diversive

Consider a typical breadwinner in search of relaxing entertainment at home after the proverbial "hard day at the office." What does "entertainment" mean? It clearly implies some form of stimulation to act as a relief from some low base-line generally designated as boredom that might characterize the person's environment, whether at the office or at home, or both. Suppose the person decides to watch television to obtain such stimulation. He twiddles the dials, until he finds a program to suit his tastes and preferences. This would be a clear instance of diversive exploration: activity that entails a search for stimuli that will result in some optimal level of stimulation, one that maintains the individual in an appropriate state of arousal. To judge by the viewing habits of the general public, it appears that the nature of the stimulation most often selected will fall at some intermediary value with regard to the uncertainty contained within it, whether considered in terms of novelty, diversity, surprisingness, or whatever. The person will be likely to eschew something thrice familiar (e.g., a rerun), while similarly avoiding something high in novelty or complexity (e.g., a program of modern art). The general principle, in other words, is that diversive exploration, aimed at establishing some optimal level of arousal in the individual, is typically satisfied by some intermediary level of uncertainty in the stimulus, which may in fact be relatively low. This is the major difference between this type of exploration and the specific, which applies when an individual is confronted with a particular stimulus generating conflict or uncertainty. In this case, the amount of effort expended in investigating or exploring the stimulus in question is postulated to be a monotonic function of its uncertainty, based on any of the above-mentioned collative variables.

Note that arousal is implicated in both of these exploratory modes, but in the one case (diversive exploration) the activity is directed at an increase in arousal level beyond a low level, while in the other (specific) it is instigated by a high level of arousal, i.e., a state of conflict or uncertainty, and directed at lowering this level.*

If it seems clear that specific exploration is not typically related to aesthetic satisfaction in any simple manner—i.e., conflict-laden stimuli are not generally considered beautiful—what about diversive? Are stimuli that are particularly effective in dispelling boredom necessarily high in aesthetic value? The previous reference to the selection of TV programs itself suggests the need for caution in this respect. Yet, certain findings suggest that stimuli that are high in aesthetic appeal are likely to meet the necessary conditions for diversive arousal.

Berlyne (1963) has approached the study of diversive exploration via the "exploratory choice" method, which involves presenting subjects with pairs of stimuli, and then asking them which of the two they wish to see (or hear) again. In general, following very brief initial exposure, subjects chose the member of the pair that was more complex, more irregular, etc., i.e., the stimulus that was higher in collative attributes, while following a longer period of initial exposure the direction of the choices was reversed. (Cf. Berlyne, 1974, chap. 14, for a review of additional similar evidence.)

Thus, it appears that once curiosity concerning a stimulus has been satisfied, the individual will prefer stimuli of lesser complexity or greater regularity. Berlyne's use of dichotomized stimulus pairs leaves open, however, the question of the optimal amount of complexity required to satisfy such diversive exploration. Subsequent research

---

* The matter is complicated by the phenomenon of the "arousal jag," which involves the seeking-out of an experience that exposes the individual to a relatively brief period of high-intensity stimulation, resulting in a marked heightening of arousal, with a subsequent sharp drop, as exemplified in many amusement-park rides, as well as thrill-seeking behavior such as sports-parachute jumping and the like. But this phenomenon, besides being subject to very marked individual differences, is sufficiently separate from the aesthetic domain that we need not swell on it.

by Crozier (1974), along with other evidence reviewed by Berlyne (1974, chap. 14), does suggest that this optimum corresponds to some intermediary level of complexity or diversity, though as in so many domains of behavior we are clearly dealing here with a multidetermined phenomenon.

Admittedly, exploratory choice can hardly be said to represent a direct expression of the perceived aesthetic value of a stimulus, Yet, it is noteworthy that the stimulus correlates of exploratory choice appear to correspond very closely with those uncovered in research in which semantic judgments of beauty or pleasingness have been used (cf. Berlyne, 1963; 1974, chaps. 5, 10). This work has established rather conclusively that hedonic value, or aesthetic satisfaction, is related to stimulus uncertainty according to an inverted-U shaped function that is very similar to that uncovered in exploratory choice studies of diversive exploration, such as Berlyne's (1971, chap. 5). More extensive factor-analytic research in this area has corroborated, as well as deepened this view of the structure of aesthetic judgment, and its relationship to arousal, and to specific and diversive exploration. Specifically, the work of Crozier (1974) and Berlyne (1974, chap. 5), and other visual and auditory research (cf. Berlyne, 1974, chap. 14) utilizing semantic-differential judgments of both artificial and artistic stimuli consistently reveals two factors, corresponding to Osgood's activity and evaluative factors. The first of these, labeled "uncertainty– arousal," is related monotonically to stimulus uncertainty and usually loads on judgments of interestingness; it is also closely correlated with measures of looking time (i.e., specific exploration). The second, labeled "hedonic tone," relates in inverted U-shaped fashion to uncertainty; it loads on scales of pleasantness, beauty, and similar dimensions, and thus appears identifiable with diversive exploration. It is Berlyne's contention (e.g., Berlyne, 1967; 1974, chap. 10), that aesthetic appreciation of stimuli involves a combination of both of these factors, i.e., the joint action of two mechanisms acting in concert, one an arousal-reducing mechanism activated by stimuli relatively high in uncertainty, the other an arousal-increasing mechanism directed at stimuli that are intermediary in uncertainty, the net result being perhaps a situation such as represented in Figure 1.

This two-factor formulation certainly fails to do justice to the complexity of aesthetic judgments, whether taken with reference to art, or to environmental stimuli. Quite apart from the adequacy of the concepts invoked to deal with the structural aspects of aesthetically significant stimuli, which will be considered at the conclusion of the section to follow, after the empirical research relevant to the formula-

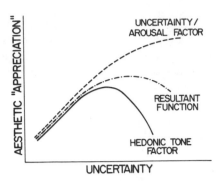

Figure 1. Hypothetical functions relating uncertainty to aesthetic response, as a composite of uncertainty/arousal and hedonic tone aspects of the response.

tion has been reviewed, it must be noted that it places a very heavy burden on the role of structure, to the exclusion of content. In particular, questions of meaning, of associations, symbolic or otherwise, formed to specific stimuli or types of stimuli are entirely left out of account. Clearly, a complete theory of environmental aesthetics will have to come to grips with this question; it hardly seems likely that the effect on the senses of an urban panorama such as the city of Jerusalem, or of an automobile graveyard along a highway, can be analyzed in exclusively structural terms.

Thus, Berlyne's theory is useful perhaps as a first approximation to the phenomena of environmental aesthetics, but one that is nevertheless of direct relevance in this context, as shown in the frequent reference on the part of writers on architecture and design to the concepts of complexity and ambiguity (Rapoport and Kantor, 1967; Rapoport and Hawkes, 1970; Lozano, 1974; Venturi, 1966) and surprisingness (e.g., Nairn, 1963, chap. 3), as qualities enhancing the aesthetic appeal of environments.

REVIEW OF ENVIRONMENTAL AESTHETICS RESEARCH

A major reason for introducing the theory of Berlyne at the outset of this paper is not only the fact that it is one of the dominant theories in the field of psychological aesthetics, but also that it appears particularly suited for application in environmental psychology, because of its stress on clearly defined stimulus parameters that have direct applicability to the physical environment. For this reason the review of the literature to follow will focus on the admittedly limited amount of research either inspired by or relevant to this theory, and the postulated role of collative variables in particular.

RESEARCH ON THE ROLE OF BERLYNE'S COLLATIVE PROPERTIES: COMPLEXITY

Of the several collative variables listed above, the one that has been most frequently utilized in environmental-aesthetics research is clearly that of complexity, and more particularly that of the aspect of complexity that we have designated as diversity. The role of this variable is incisively brought out in an experimental study by Schwarz and Werbik (1971), in which toy-scaled models of streets were constructed with the houses along the street varying along two dimensions: distance from the street, and angle or orientation relative to the street. Combinations of three distances and four angles were utilized, so as to result in patterns containing one of four specifiable levels of uncertainty, based on the number of different types of houses included. Movies simulating a trip along this street were then shown to subjects with instructions to rate them along several semantic-differential scales that referred to attributes that had been shown to load either positively or negatively on Osgood's evaluative and activity factors. The results showed that those mood-adjectives that were positively related to the evaluative factor (e.g., gay, harmonious, satisfying) varied with uncertainty according to an inverted-U shaped function, while the opposite was true for moods inversely related to this factor (e.g., somber, sad, serious, dark). The same pattern of results was obtained with respect to moods loading on the activity factor.

While there will undoubtedly be those who will scoff at the artificial nature of the stimulus materials employed in this study as a simulation of real environments, the study remains as a model of the way in which semantic-differential type of judgments can be fruitfully employed in this field of research, i.e., by relating them systematically to independently specified stimulus parameters, whether these be experimentally manipulated, or assessed indirectly, with respect to actual ready-made environmental scenes.

An example of the latter approach is a study of the writer's (Wohlwill, 1968), in which color slides of a wide variety of environmental settings were scaled with respect to diversity, with reference to specified attributes such as color, texture, direction of dominant lines in the field, and shapes of dominant elements in the field. These ratings were made by a set of independent judges carefully instructed in these judgments, and slides that showed undue dispersion for these ratings were discarded. Those remaining were ordered in terms of seven levels of composite diversity, with two scenes at each level, and given to college subjects, first with instructions to look at them as often as they wished, under .5-second exposure, and subsequently with instruction to rate the scenes for degree of liking. In accordance with the above-mentioned distinction between the two factors of uncertainty arousal and hedonic tone, two contrasting functions were obtained, the subjects' looking times increasing in roughly monotonic fashion with diversity, whereas a very irregularly shaped inverted-U function appeared to emerge from the ratings of liking. Much more dramatic evidence for such an inverted-U function was obtained subsequently when the same slides were employed in a study of preferential choice by the paired-comparison method, with children between 6 and 14 years of age (Wohlwill, 1975). Again voluntary exposure time provided a much more nearly monotonically increasing function.

One difficulty with the series of stimuli utilized in the research just referred to is that the environmental scenes scaled for complexity ranged from pictures of an Arctic landscape to those of industrial and commercial sections of urban areas. Kaplan, Kaplan, and Wendt (1972) noted that complexity thus appeared to be contaminated with the differentiation between the natural and the man-made world, which they were inclined to consider the factor actually determining the obtained results, particularly the liking data. They proceeded to obtain judgments of liking for unselected scenes representing these two types of environments that seemingly bore out their interpretation, showing a large overall difference in the ratings favoring their nature set.

It should be noted that complexity, as measured in their case by asking the same subjects who made the preference judgments for direct ratings of complexity, still turned out to be correlated with preference within each of the three stimulus sets that Kaplan *et al.* employed (including a set made up of natural and man-made elements). In fact, preference appeared to increase linearly with complexity, in apparent conflict with the inverted-U shaped relationship typically encountered. It is possible, of course, that the range of the complexity variable sampled in these slides was not sufficient to reveal the decreasing phase of the curve. This possibility, along with a much more fundamental question that Kaplan and associates' criticism of the

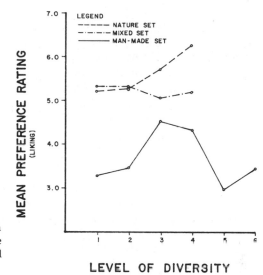

Figure 2. Relationship between stimulus diversity and preference ratings, for nature, man-made, and mixed slide sets.

writer's research brings up, led to a further study of this problem carried out at the Penn State laboratories.*

This study started with the selection of three parallel series of slides of environmental stimuli, one consisting of pure nature scenes, a second of pure man-made scenes, and a third involving a mixture of the two types in juxtaposition. Within each series the stimuli were scaled for complexity, in the same manner as indicated above for the previous studies, i.e., based on the slides' composite diversity, along preselected dimensions. This scaling led to the first important finding, which became evident before the data-collection phase even began: We found it impossible to locate natural scenes of a degree of diversity comparable to those from the man-made environment falling at the upper end of this scale. This observation suggests, although it does not prove, that there is an intrinsic relationship between the nature versus man-made differentiation and the complexity variable, which was presumably implicated in the seeming confounding of these two variables in the earlier study.

Turning now to the results, two kinds of ratings were obtained for these slides, interestingness and liking, but since the results for both were very similar, only the latter will be considered here. As Figure 2

_____

* This study was carried out as part of an environmental psychology laboratory, with the collaboration of Harry Heft, Fred Hurand, Robert LeDoux, Elliot Reiff, and Richard Titus. The results of this research are reported in greater detail in Wohlwill (1976).

shows, for the man-made set the expected inverted-U shaped relationship of liking to complexity does obtain, while over the more limited
range of this variable represented in the nature set the relationship is
more nearly monotonic. (As for the man/nature set, the ratings were not
consistently related to complexity at all.) Finally, Kaplan and associates' suggestion that the major variable of importance is the differentiation between the natural and the man-made world received some
confirmation in our data, as shown by the consistent and highly
significant difference in overall level between the latter and the other
two sets, even when only the first four levels of the diversity variable
are considered. This finding clearly leaves unanswered the question as
to the basis for and stimulus correlates of this differentiation. While in
our case sampling of the slides in the different categories might have
played a part (the natural environment scenes were made up
predominantly of travel slides, while those of the man-made came
from a greater variety of sources, and were thus less strongly biased in
the direction of visual appeal), this artifact does not seem applicable to
the similar pronature bias found by Kaplan *et al.* (1972).

The preceding studies appear to be the only ones utilizing
environmental stimuli in which independently defined measures of
complexity, or any of Berlyne's collative variables, for that matter,
have been related to behavioral responses or evaluative judgments.
Even the complexity variable appears to have been studied only in the
diversity sense, despite the clear relevance of the structural aspect of
complexity to the individual's evaluation of and behavior in the
environment; cf. Lynch's (1960) "imageability" construct, for instance.* Although Lynch himself stressed the affective relevance of
imageability of urban environments, and common sense itself suggests that structural aspects of the spatial layout of a city or town
would have a very direct bearing on the aesthetic satisfaction (or lack
of it) to be derived from it, this question has not been directly
confronted in research.

A rather large number of studies have, however, employed
ratings, either of complexity per se, or of scales related to this variable
in either of the two above-mentioned senses, in the context of

---

* Complexity in this structural sense is in fact being investigated in an interesting
experimental study at the Technion in Haifa, Israel, by Yigal Tzamir, who is
simulating trips through artificial urban environments constructed so as to vary
particular structural features by means of an environmental simulator, and studying
orientation and way-finding behavior as a function of these features. While specifically directed at Lynch's imageability concept, this experimental approach is closely
akin to that of Schwarz and Werbik (1971) referred to earlier.

correlational or factor-analytic studies of the semantic space represented by particular environments. Included here are studies of rooms (Küller, 1972), buildings (Canter, 1969), shopping centers (Downs, 1970), cities (Lowenthal and Riel, 1972b), landscapes (Küller, 1972), and natural recreation areas (Calvin, Dearinger, and Curtin, 1972), as well as of unselected environmental scenes (Küller, 1972).

From the above studies a rather consistent finding emerges: Rated complexity is generally uncorrelated with ratings of pleasingness or beauty, or does not load on the same factor. (Not all of the studies report the intercorrelation matrices.) Two noteworthy exceptions to this statement may be mentioned: Küller (1972) found that for ratings of a set of 15 slides that were partly of indoor and partly of outdoor scenes, both "complex" and "uncomplicated" loaded on the same factor as pleasantness and beauty; in the case of a set of slides of interiors, on the other hand, "simple" loaded most strongly on a "social status" factor. Further, Lowenthal and Riel (1972b), in their study of ratings based on paths walked by the subject through urban areas, employed two scales that correspond roughly to the two varieties of complexity differentiated above (they did not include a complexity scale per se). "Ordered–chaotic," presumably referring to structural complexity, was highly correlated with beautiful–ugly (+.63); on the other hand, "contrasting–uniform," which more nearly represents the diversity variable, was uncorrelated with beautiful–ugly.

In interpreting the above results, and particularly the predominant lack of correlation between the complexity and the beauty or pleasantness ratings, two major points must be borne in mind. First, little if any attempt was made to define these scales for the subjects, so that the ratings represent subjective impressions whose relationship to specifiable characteristics of the environments concerned must remain moot; for the most part, the investigators cited were not interested in studying functional relationships between environmental characteristics and affective responses, in the sense of the research by the writer and Kaplan et al. described earlier. Second, the correlation coefficient is of course predicated on the existence of a linear relationship; if, as there is both theoretical and empirical reason to believe, the relationship between complexity and affect is in fact curvilinear, it would not be effectively revealed in these data, with a correlation coefficient of low magnitude resulting. At the same time, the divergent results noted above for Lowenthal and Riel for the ordered–chaotic versus the contrasting–uniform scales not only provide further support for the importance of the distinction between the two senses of complexity, but suggest that structural complexity may indeed be more nearly linearly related to aesthetic evaluation, though in a negative sense. It should be noted, however, that "ordered" does not necessarily mean lacking in complexity, i.e., devoid of information, but rather a degree of redundancy imposed on a configuration that in terms of number and/or diversity of elements may in fact be relatively high in potential amount of information.

To sum up, it is apparent that complexity, however much it might be touted by aestheticians as well as designers, plays an uncertain role at best in the individual's aesthetic response to the environment. While more refined scaling may verify the suggestion from some of the research thus far that some intermediate degree of complexity is optimally conductive to aesthetic satisfaction, it is also apt to reveal—providing we undertake the kind of research that will permit it to do so—that there are important factors relating to the individual, as well as to the temporal and spatial context, that modulate the degree of arousal that will be found satisfying, and thus the amount of complexity of stimulation considered optimal. Indeed, there is impressive empirical evidence concerning the extent of individual differences in preference for complexity (cf. Vitz, 1966), suggesting in fact that the optimization principle may be itself related to the level of adaptation with respect to environmental complexity that an individual has built up in his or her prior experience (cf. Wohlwill, 1974). And apart from interindividual differences, there are reasons to anticipate major fluctuations in preferred arousal levels across the diurnal cycle, as well as different situational contexts: a point stressed in Fiske and Maddi's (1961) theoretical account of the function of stimulus variation, while Lozano (1974) has emphasized it in relation to the characteristic of optimal aesthetic value of environments, from the perspective of the urban designer and planner.

Apart from this issue of inter- and intraindividual variation in preferred levels of complexity, the question of the dimensionality of this attribute is one that will need to be faced. There are good reasons to distrust a unidimensional conception of complexity, and to suspect that complexity in the sense of diversity may well be altogether unrelated to the structural aspect of this construct. To deal effectively with the latter, however, we shall have to devise ways of operationalizing order, possibly in terms of the redundancy contained within a configuration of elements, and to examine its role in interaction with information content in the diversity sense (cf. Mindus, 1968). But it is clear that no valuable insights into these questions can possibly be expected from further mindless proliferation of factor analyses of semantic-differential ratings.

THE ROLE OF OTHER COLLATIVE VARIABLES

In considering the role of such variables as novelty, incongruity, ambiguity, and surprisingness, we have little evidence based on independent manipulation or even objective assessment of these attributes, though their potent role in motivating exploration of the environment can be inferred from the tourist trade and similar manifestations of aggregate behavior, as noted earlier. We do, however, have more incidental information relating to some of these from semantic-differential research, as well as rather more direct evidence touching on incongruity and ambiguity. Let us review this evidence for each of the above-mentioned variables.

Novelty

A scale clearly relating to this attribute, "commonplace–unique," was used in the studies of Canter (1969) on exteriors, and Calvin et al. on natural landscapes. In Canter's study, architects' ratings on this scale loaded heavily on the first factor (i.e., identified as "character," but representing basically the evaluative factor); curiously, for nonarchitects, this scale appeared to define a unique factor, and did not load on either of the primary ones. In the case of Calvin and associates' study, on the other hand, the ratings on this scale loaded very highly (.89) on the evaluative factor; more particularly they correlated .93 with ratings of beauty. Finally, Küller (1972) used several different variables related to the novelty concept, with inconsistent results: In his ratings of interiors, "ordinary" appeared to define an originality factor rather than loading on the evaluative factor, while in the ratings of the mixed landscape and interiors set "original," "unfamiliar," and "odd" all loaded most strongly on the first, evaluative factor but positively for "ordinary," and negatively for the other two. (A further related attribute, "unusual," loaded only to a much more moderate extent on the second, potency factor).*

* It should be noted that Küller's scales were all defined in unipolar terms, i.e., designated by a single term such as "quiet," "untidy," etc.; the subjects were asked to rate the *amount* of the attribute contained in each slide in terms of a scale from "least" to "much." Given the obviously bipolar character of most of the attributes referred to in these scales, this procedure appears inappropriate.

The above findings are somewhat puzzling. To begin with, one might expect a pattern for novelty paralleling that for complexity, i.e., a monotonically increasing relationship to exploratory behavior, or interest, but one approaching instead the inverted-U shaped function in relation to liking, beauty, etc. (e.g., Berlyne, 1974, chap. 8). One suspects that the results actually obtained are in part related to the sampling of environments included; at least in the case of Calvin *et al.* and Küller, none were used that one would consider particularly foreign to their subjects' environmental experience (Canter did not reproduce his stimuli in the published article). On the other hand, the finding by Küller that "odd" and "unfamiliar" loaded negatively (and rather highly so) on the evaluative factor suggests that novelty, like complexity, may be a multidimensional attribute. Much would depend, however, on possible artifacts of the factor-analytic solution employed here; it is unfortunate that the original correlation matrices are not included in Küller's monograph, as they might have helped to clarify the meaning of his results.

In the light of the available data, one is impelled to withhold judgment about the actual role of the novelty variable in aesthetic evaluation of environments, all the more since no attempt has been made to obtain independent measures of this attribute. The attempt of Leopold (1971; cf. also Leopold and Marchant, 1968) to construct a uniqueness ratio for the assessment of the aesthetic quality of riverbed scenery, for use in environmental-impact assessment and in testimony over conservation-related issues (e.g., relating to proposals for dam construction) represents one effort in this direction, but has not been applied to empirical work on aesthetic judgment. The assumption behind the use of this index, however, is clearly that uniqueness is a measure of aesthetic quality. At present that assumption must be taken on faith, however plausible it might appear to some, though the results of Calvin *et al.*, obtained with pictorial material similar to the type of natural landscape Leopold has been concerned about, lend it a certain measure of credibility.

Surprisingness

Of the semantic-differential studies, only Küller used a scale referring to this attribute directly, and only with the mixed interior and landscape set. His results show only a weak loading on the first evaluative factor for this scale; it was mainly related to the same factor as the "unusual" scale, co-defining a factor identified as originality, even though the "original" scale itself did not load on it as strongly.

The variable is deserving of more explicit attention on the part of environmental psychologists, all the more since the arousal value of violations of expectancies occupies a central place in several theoretical formulations of aesthetics, e.g., Meyer (1956) for music, Koestler (1949) for literary arts and the stage. Its systematic investigation in the field of environment and behavior would, however, require that a temporal aspect be built into our environmental-perception methodology, which has been notably absent in this area, with its almost exclusive resort to photographic slides. Lowenthal and Riel's work (1972b) based on walks taken through sections of different cities represents a notable exception to this rule, but the scales employed unfortunately failed to exploit this temporal feature.

Ambiguity or Uncertainty

None of the semantic-differential studies appear to have included scales specifically relating to this attribute. It is, however, relevant to the characteristic of "mystery," which the Kaplans, on the basis of several studies (S. Kaplan, 1975; R. Kaplan, 1973, 1975) have established as an important determinant of preference for both urban and natural scenes. This variable was defined for the subjects rating it as "the promise of further information based on a change in the vantage point of the observer. Consider whether you would learn more if you could walk deeper into the scene" (R. Kaplan, 1975). It is exemplified by pictures of roads curving out of the frame of the picture, or disappearing behind a building, or the like. It thus involves the generation of uncertainty, though in a sense quite different from that involved in the informational calculus applied to the complexity of a set of elements. At the same time the concept of "promise of further information" in reference to this characteristic clearly and directly relates it to Berlyne's collative properties. Thus, the finding that mystery correlated to the extent of .64 with ratings of pleasingness and liking (R. Kaplan, 1975), is impressive. Indeed, the correlation increased to .86 when another variable to be considered presently, i.e., that of coherence, was partialled out. This strong relationship appears, however, somewhat inconsistent with the relationship of uncertainty to preference found in the complexity research. The emphasis on mystery is, furthermore, at variance with the equally strong emphasis by S. Kaplan (1975) on the variable of legibility, which, at least in a connotative sense, appears to refer to the reduction of uncertainty, rather than its generation. We will return to this issue in the appraisal of the role of uncertainty following this literature review; it is clearly of importance for environmental aesthetics, as for aesthetics generally.

Incongruity and Congruence

Perhaps nowhere are the difficulties of an attempt to apply Berlyne's theory of collative variables to environmental aesthetics brought out more sharply than in the case of the incongruity attribute. This characteristic refers to the juxtaposition of elements that are in some way difficult to reconcile or discordant, or which involves the placement of a stimulus in a context foreign to it. It is thus reasonable that Berlyne should conceive of it as a source of conflict, and thus of exploratory activity aimed at resolving or at least reducing it. There is indeed ample evidence that incongruity does enhance such activity, as shown in increased length of voluntary exploration of stimuli such as a dog with feathers and the head of a bird, or a cow with airplane tails (Berlyne, 1958; Nunnally, Faw, and Bashford, 1969). These are, of course, measures of specific exploration, and tell us little concerning the individual's affective response to the stimuli, as noted earlier. Incongruity was not, unfortunately, among the variables included in Berlyne's more recent factor-analytic research based on verbal ratings that led to the two-factorial model cited in the section on complexity, so that we do not know the extent or direction of the loading of this variable on the hedonic tone factor. There is, however, one study by Berlyne (1963), referred to earlier in connection with the specific–diversive distinction, that reports pleasingness and interestingness ratings for the standard series of stimulus pairs that Berlyne has used to study effects of incongruity. The mean pleasantness ratings for the incongruous stimuli not only were much lower than for the congruous members of the pairs (2.6 versus 4.8 on a 7-point scale), but fell well below the means for any of the other categories of stimuli employed in the same study (e.g., involving irregularity, heterogeneity, asymmetry, etc., illustrated via abstract designs).*

This finding appears to have its clear counterpart in the realm of environmental aesthetics. There is no question that at least certain

---

* Significantly, in the exploratory-choice portion of the study (cf. above), the choices did not differ between the brief and long conditions of prior exposure to the stimulus pair, contrary to the findings for the variables of diversity, irregularity, and patterning. In other words, not only in terms of pleasingness, but exploratory activity as well, a little incongruity "goes a long way," i.e., after the brief exposure time the subjects evinced no more interest in further exploration of the incongruous stimuli than after longer exposure.

highly incongruous elements in our environment manage to attract attention and curiosity: The popularity of Lake Havasu City, with its display of London Bridge planted in the middle of the Arizona desert, which at last reports has been rivaling the Grand Canyon as a tourist attraction, is a good case in point. Yet, it appears highly doubtful that incongruity exerts similar positive effects on the individual's affective response to environmental stimuli.

Indeed, incongruity has become a chief source of public concern in the controversies over environmental issues, as shown in the campaigns to ban billboards, to zone areas against commercial or industrial use, to keep highways out of urban parks, and to ensure uniformity of style in the design of residential developments; it has similarly become a potent issue in litigation involving questions of aesthetics, as we will note in the concluding section of this paper. It may thus be expected that incongruity will show strongly negative loadings on any hedonic tone or evaluative factor and be inversely related to ratings on scales such as beautiful–ugly. What evidence can be adduced in support of this hypothesis?

The evidence at this juncture is admittedly scattered and inconclusive. To begin with, it is difficult to investigate the role of incongruity in any generic sense, for two reasons. First, what is or is not experienced as incongruous is necessarily determined by an individual's experience, and conceptions as to what type of sights are or are not appropriate in some particular context, or in juxtaposition to other sights. For example, the sound of a baseball game encountered along a trail through a wilderness area may appear highly incongruous to the hiker in search of a wilderness experience, but much less so to a mining prospector, for whom the area has mainly a utilitarian interest. Thus, incongruity, probably more so than other collative variables, should be studied relative to well-defined normative groups sharing a common set of expectations. Second, incongruity can arise in quite diverse ways, which are probably not at all equivalent. It may result, first of all, from a clash of styles, such as most older cities (cf. the center of postwar Rotterdam) or campuses inevitably display in the appearance of their buildings. Alternatively, it may arise from marked disparity of scale, color, or shape between a particular building and its surroundings, as in the case of a single skyscraper towering over an area of sedate one- or two-story buildings, or the spanking white of a civic center such as New York's Lincoln Center amid a gray, drab, decaying slum.

Considering how frequently both of these sources of incongruity develop into the object of major controversies in which architectural

designers become caught up, and the virtual exploitation of incongruity of scale and shape as an architectural device in the work of at least
one noted contemporary designer, Robert Venturi, it is remarkable
how little systematic attention they have received from researchers.
One suspects that there are considerable variations from one individual to another in tolerance for or appreciation of salient disparities of
style, from the person who finds such clashes stimulating and enhancing the interest of a campus or urban area to those who insist on a
maximal degree of blending-in of the new with the old, and a
minimizing of contrasts.

One of the rating scales used by Lowenthal and Riel (1972b) is
perhaps relevant in this context. This is the contrasting–uniform scale
cited above in connection with the effects of diversity. As noted
earlier, the ratings on that scale did not correlate with the beautiful–
ugly dimension, or, for that matter, with any other of 25 scales used in
this investigation! Küller (1972) used a scale defined by the term "of
pure style," which appears also pertinent; for his set of slides of
interiors, this attribute likewise failed to load on the evaluative factor,
but instead appeared to co-determine a "unity" factor. While this
factor also emerged from the ratings of the mixed interior–exterior
scenes, the of-pure-style ratings loaded equally strongly on the evaluative factor in this case. The highly heterogeneous materials used make
it difficult to determine the basis for this finding, e.g., whether it may
reflect response to architectural style, to clashes between man-made
and natural settings, or what.

Fittingness and Compatibility

A somewhat special case of the role of incongruity in the second
of the two senses just cited concerns the perceived fittingness or lack
of it of particular man-made structures in a particular natural setting.
This is clearly in part an aesthetic question; perhaps because of its
relevance to the land-use problem, it has attracted a certain amount of
attention in research.

Steinitz and Way (1969) have presented a model for the analysis of
the visual impact of man-made changes introduced in a landscape (in
the form of residential and industrial buildings, as well as commercial
developments on a larger scale), by measuring the perceived *difference*
between landscapes containing those changes, and equivalent control
landscapes, and relating these perceptual difference values to a variety
of other data, some relating to objective topographical features (e.g.,

opacity of vegetation; ratio of area of development to total area), some to perceptual attributes of the landscape (e.g., ratings of "naturalness" of the landscape), and some to affective measures (e.g., ratings of liking of the structure or use of the land represented by the change). Visual impact, operationalized in the manner indicated, was found to be highly related to three physical attributes, i.e., proportion of area covered by the development, opacity of vegetation, and extent of color contrast.

Steinitz and Way's approach appears promising in its conception of visual impact and in the methodology employed to assess it, but it does not seem to have been more generally applied. The same problem has been dealt with somewhat more extensively by Hendrix and Fabos (1975), who have devised an "index of compatibility." This index is based on a set of 11 categories of land uses (such as wetlands, recreation, commercial, etc.) represented photographically and juxta-posed pair-wise, according to a paired-comparison procedure, with instructions to observers to judge the extent to which each type of landuse would affect the "visual quality" of the one it was paired with, from positively (enhancing) to negatively (detrimental). This yielded a measure of compatibility for any particular adjacency, i.e., pairing of land-use types. (Thus, the index was in the "high compati-bility" range for such pairings as forest-wetlands and open water-recreation; it was in the lowest range of such pairings as commercial-forest, and recreation-transportation.) Zube, Pitt, and Anderson (1974) employed this index (among a wealth of other measures; cf. Zube's chapter in this volume) in their comprehensive study of perceived scenic resources in the Connecticut River valley, and found that this measure of land-use compatibility (with each adjacency weighed by the extent of the boundary between them in the visual field) accounted for 27.5% of the variance in judged scenic resource value, more than twice as much as the second highest dimension, "absolute relative relief."

Finally, similar results appear to be emerging from a research project of the writer's currently under way on the evaluation of an urban waterfront area. This study was carried out on a set of eight sites along an urban waterfront, each site being sampled through five different views. Evaluative judgments were obtained under both field and slide conditions, as well as with subjects familiar as well as unfamiliar with the scenes evaluated. These judgments were corre-lated with independently derived ratings of the objective environ-ments, based on two sets of measures adapted from Zube et al., one involving prominence ratings for particular features of the landscape (e.g., vegetation, highways), the other involving landscape quality ratings (e.g., open-closed, natural artificial).

Thus far only the slide data for the unfamiliar subjects have been analyzed; they indicate a very high correlation across views between the two evaluative scales of harmonious clashing and beautiful–ugly, admittedly a less than startling result. The results show further that the evaluative ratings of congruous–incongruous and obtrusive–unobtrusive, both referring to the juxtaposition of man-made and natural elements in the scene, were themselves highly correlated with judgments of beauty, as well as with the independently derived ratings of clean–dirty, unified–chaotic, and natural-artificial. The last-mentioned finding is particulary noteworthy, since it indicates that a predominance of the natural over the man-made appears to be essential in order for the interplay between the two kinds of elements to be rated as congruous. Perhaps we see here, in a different guise, another manifestation of the pronature bias to which Kaplan *et al.* (1972) pointed, as noted in the discussion of the role of complexity.

The Place of Uncertainty in Environmental Aesthetics: A Reappraisal

The partial review of the literature on variables such as complexity, novelty, congruity–incongruity permits the generalization that Berlyne's collative variables do relate positively to attention, interest, and possibly arousal (the picture is a conflicting one in regard to the latter, in part due to the paucity of adequate direct measures of arousal). The relationship of these same variables to hedonic tone, or measures of affect, aesthetic evaluation, preference, and the like is much more complex, however. Such responses appear to be maximal either for intermediary levels of these variables, or for combinations of stimulus parameters involving certain organizational properties that have been variously designated as balance, harmony, coherence, unity, legibility, congruence, fittingness, etc. (e.g., Ertel, 1973; S. Kaplan, 1975). These properties, far from involving a maximization of uncertainty, appear to keep the latter within very definite bounds, though without necessarily causing it to dissipate altogether.

The critical point with respect to the role of these Gestaltist-flavored variables is that they refer predominantly to structural properties of the configuration of a visual field, and of the elements of the field in relation to one another. Such terms as harmony, balance, unity, coherence, fittingness are all relational, and presuppose a structural approach to the assessment of these variables. This has been largely absent in this field, though to a limited extent the work on visual impact by Steinitz and Way (1969) and the land-use compatibility index of Hendrix and Fabos (1974) referred to above do entail such a structural analysis of the visual field. Admittedly, when dealing with real, ready-made environments (as opposed to artificially constructed stimuli), this type of analysis becomes difficult to carry out, and one can understand the readiness of investigators to resort to the shortcut of direct verbal ratings in handling such variables as harmony, unity, and coherence.

At the same time efforts aimed at experimentally manipulating these structural properties through the use of artificially constructed simulations of actual environments would seem to be called for (cf. the work of Tzamir referred to earlier). These should help to clarify the role of redundancies of different kinds, both structural (e.g., arrangement of streets in a grid pattern) and associational (e.g., boats on a lake) in environmental aesthetics.

In this connection it is interesting to note Sanoff's (1969) observation in regard to the place of the "good Gestalt" in environmental design. As Sanoff points out, good Gestalts entail a high degree of internal redundancy; on this basis Sanoff is inclined to rule out this principle as a basis for design, "because it tends towards states of diminishing novelty in the environment." The question, of course, is just how much novelty (or uncertainty) the individual actually demands in his everday world. Here again we are presumably facing an optimization problem: the proliferation of perfect circles (the case par excellence of the good Gestalt), found in some suburban developments may well engender boredom, but it seems clear that a mazelike pattern of streets devoid of any discernible order would prove even less viable or satisfying. It is worth referring in this connection to an experimental finding by Mindus (1968), albeit with artificial sequences of individual tones: Redundancy may interact with information, such that it actually increases interest when introduced into a high-information (8-tone) sequence, whereas it engenders boredom when introduced into a low-information (4-tone) sequence.

Lozano (1974) has recently contributed a fresh perspective on this question in his argument for the need for variation across space as well as time in amount of order, information, etc., in the visual environment. While little direct evidence on the role of such variation is available in the environmental literature, it is consistent with psychological theories of the role of variation in arousal level, notably that of Fiske and Maddi (1961). At the same time the seeming contradiction between the Kaplans' emphasis on legibility and mystery may be resolved (S. Kaplan, personal communication), if we admit the possibility that the occasional generation of uncertainty in a temporal sense, as implied in the definition of mystery as "the promise of further information," may be positively valued, within the context of an environment that is legible, i.e., affords clarity in the integration of the elements simultaneously present to view.

ISSUES OF METHODOLOGY IN THE STUDY OF ENVIRONMENTAL AESTHETICS

The field of environmental aesthetics can, in many ways, be taken as representative of the environment-and-behavior field, with respect to the difficult questions of methodology with which it confronts the investigator. No thorough treatment of this question is possible within the limits of this paper, but some of the major problems warrant brief discussion. They will be considered under four different aspects.

DEFINITION AND MEASUREMENT OF ENVIRONMENTAL VARIABLES

One of the most difficult problems to be resolved in this field concerns the specification and measurement of the properties or attributes of the environment that are chosen for investigation. A large amount of the research on so-called environmental perception has completely sidestepped this problem, by the expedient of selecting environmental sites, views, structures, paths, or verbally designated locales or regions, without any attempt to assess these with respect to specified variables of the stimulus. The preferred approach has been to present such frequently arbitrarily chosen environmental stimuli (usually simulated through photographic slides) to subjects with instructions to rate them on a large number of verbal rating scales (generally termed "semantic-differential") through which the individual's perception or evaluation of the stimuli might be assessed. The outcome of such studies has inevitably been limited to correlation matrices indicating the intercorrelations among these various ratings, which are then reducible via factor analysis to a smaller number of dimensions. The information derived from such analysis is, however, purely descriptive; moreover, it describes the manner in which subjects use the verbal scales in response to environmental stimuli of a given type, without telling us anything about the role played in these judgments by any specifiable environmental characteristics.*

What is lacking in such studies, in other words, is the possibility for establishing functional relationships among variables, which, one would suppose, is the business of a scientifically based research effort, however the variables and the relationship between them may be construed. Such relationships can only come into play once the environmental stimuli selected for study are in some way assessed with respect to specified characteristics, independently of the evaluative or behavioral responses to be made to them.

Several approaches are available to the investigator who is interested in following this prescription. They appear to fall into two major types: the physical and the judgmental.

Physical Measurement Approaches

For certain purposes, it is possible to obtain objective measures of particular physical characteristics of the environment that may pertinently be related to psychological measures. This is true particularly for intensive dimensions, such as temperature, noise level, and, in the visual domain, brightness, though in the latter case particularly some way would need to be found to integrate the values of this scale over the total visual field, as can fairly readily be done for photographic slides or prints. Unfortunately, the variables most relevant for aesthetics are apt not to be these intensive dimensions, but rather more qualitative ones, relating to structural and organizational properties, as noted above. Here little progress has been made thus far, though advances in the field of computer-analysis of visual stimuli (e.g., Wathen-Dunn, 1967, esp. pp. 354–412) may prove helpful.

The situation is obviously simplified if the environmental attributes refer, not to highly abstracted dimensions such as complexity, but rather to specific concrete features of an environment selected for

---

* Relevant in this connection is the important control study carried out by Lowenthal and Riel (1972b), who compared the patterns of their correlations among their response scales when these were applied to actual environmental stimuli (based on walks through urban areas) with those obtained by asking subjects simply to rate each scale against every other scale. In this manner they arrived at a semantic base-line, reflecting the connotations of the various adjectival terms in the abstract, independently of any specific environmental referent. Not surprisingly, a substantial proportion of these correlations were closely similar for the matrices, thus pointing to the indeterminate nature of the contribution of the particular environments experienced, not to the ratings themselves, certainly, but to the structure of the intercorrelations among them.

their possible aesthetic relevance. Examples are provided in the list of physical and biological and water-quality factors utilized by Leopold (1971) in his work on the assessment of "uniqueness" values for riverbed areas, and the landscape-feature checklist employed in the Connecticut River valley study of Zube *et al.* (1974) (cf. also Fabos *et al.*, 1973). In addition, topograpical dimensions of the landscape may be measured by methods derived from geographical, geomorphological, or landscape-architecture classifications; their use is again illustrated in the work of Zube *et al.*, as well as in the visual-impact model of Steinitz and Way (1969) (cf. also Litton, 1972).

One limitation of some of these latter methods is that the attributes chosen are often difficult to relate to the evaluative or behavioral measures on any priori basis, except insofar as they may be combined or interrelated to produce an index of more direct relevance on aesthetic grounds. The uniqueness index of Leopold represents a case in point, as do indices of naturalism–artificialism that could be based on the incidence of natural versus man-made elements in a scene.

A variant of the above approaches has been employed by Shafer (e.g., Shafer, Hamilton, and Schmidt, 1969), who described natural-landscape configurations on a total of 46 variables (combinations of zones, such as sky, vegetation, lake, etc., and parameters such as perimeter, area, etc., along with certain additional attributes). These physical measures were subjected to factor analysis in order to derive a smaller set of factors describing the major dimensions of these environments, which were then placed into a multiple regression equation to predict preference judgments. It would seem that multidimensional scaling methodology is even better adapted to this purpose; both, however, suffer from the rather synthetic nature of the dimensions or factors resulting from such analyses. These serve quite adequately for purely predictive or descriptive purposes, but may prove difficult to place into systematic relationship with behavioral measures in terms of any set of a priori defined hypotheses or theoretical principles.

Judgmental Methods

While even the "pure" physical methods just discussed may involve a certain degree of personal judgment (e.g., in the application of landscape-dimension categories to particular instances), the categories employed are generally defined in sufficiently concrete and objective terms that problems of judgment, such as reliability, consistency, accuracy, etc., remain fairly minor. For many purposes, however, the investigator may need to resort to human judges to assess dimensions of the environment that have an objective referent in principle, but nevertheless are not susceptible to direct measurement via physical indices. A good example is provided by the index of land-use compatibility of Hendrix and Fabos (1975) described above, as well as by the writer's assessment of complexity, in terms of composite diversity of a scene (Wohlwill, 1968). Jacobs and Way's (1969) approach to the assessment of the amount of development of a visually presented landscape falls in the same category, as do Craik's (1972) landscape-rating scales and graphic landscape typology derived from Litton's work. The landscape-feature prominence scales, as well as the landscape-quality ratings adapted from the work of Zube *et al.* (1974) in the writer's above-mentioned research evaluation of an urban waterfront, may also be cited here.

Despite the element of subjectivity in the rating scale and categorization methods that are relied upon in this approach, the reliability of the resulting values has generally been found to be acceptable, and in some cases quite high, as in the agreement among different groups of observers in the use of the Landscape Feature Checklist by Zube *et al.* (1974, cf. Table III-15; cf. also Zube's chapter in this volume). Pretraining in the use of the instruments, along with carefully worded and pretested instructions, can undoubtedly do much to increase the usefulness of this method. If it is to be employed effectively, however, two conditions must be met.

The first is that the scales should refer to actual attributes of the environment, rather than to purely subjective experience. This criterion has on some occasions been lost sight of; for example, Zube and associates' semantic landscape description and evaluation scales combined terms such as "natural–man-made" and "light–dark," which have a clear environmental referent with scales such as "pleasant–unpleasant" and "ugly–beautiful," which are clearly terms referring to the individual's affective response.* The second point is that in the interest of further differentiating the stimulus or environmental side

* Rachel Kaplan (1975) has contributed an articulate and vigorous defense of subjective or phenomenological definition of environmental variables, the goal of objectivity being, in her view, unrealistic, given the frequently low interjudge reliability in the assessment of variables not measurable through direct physical techniques. Kaplan further argues that even where particular physical variables may be objectively measurable, they may not be those of most direct relevance to preference or aesthetic judgment, because they lack psychological impact. The data of Zube *et al.*, as well as those obtained in our most recent research, would bear out this criticism only in part, but the issue is clearly one of focal importance.

from the response or behavioral side, the individuals used for the environmental assessment should not be those from whom the evaluative responses are obtained. Again, this desideratum has been violated in much of this work, as in the work of Kaplan *et al.* (1972) with complexity versus preference and R. Kaplan's (1973) research on the relationship of coherence and mystery to preference. The latter author does cite some evidence suggesting that the data are not in fact contaminated by this nonindependence of the ratings, but it is indirect at best.

SAMPLING OF ENVIRONMENTS

The problem of the selection of stimuli in environmental-aesthetics research is in its way as difficult and as much in need of systematic, explicit confrontation, as the preceding one of the definition and measurement of the stimuli. Indeed, it is closely related to the former, since the sampling of environments becomes an issue precisely because we wish to arrive at a set of stimuli that will provide information on the role of specified dimensions or attributes, without being able to subject those variables to experimental control.

The problem has generally been handled in an all too pragmatic, not to say arbitrary and haphazard fashion, by selecting a set of stimuli that is in some sense considered to represent a random sample of environments. The matter is perhaps less serious in the case of investigations that deliberately focus on a specific geographic setting, as in the case of Zube and associates' Connecticut River valley study, except for the obvious question of the generalizability of the results from such an ecologically delimited area. Yet, when we consider the alternatives open to the investigator, it appears that this may well turn out to be the most reasonable strategy.

Psychologists have by and large ignored the problem of sampling of stimuli, as opposed to that of subjects, due in large part to the dominant role in perceptual as well as aesthetic research of the model of laboratory experimentation, where stimuli are made to order to represent particular variables that the experimenter wishes to manipulate. Thus, the issue of sampling has rarely arisen in this type of work. One psychologist who consistently criticized this approach in the study of perception, and of behavior more generally, was Brunswik (1944, 1956). This author argued for ecological sampling of the environment, in order to capture the broad range of conditions that actually confronted an individual, as opposed to the artificial varying of single variables, all else being held constant, that is, the earmark of

the laboratory. The term "representative" was never fully operationalized by Brunswik, however, and the methodological difficulties raised
by it were not thought through very extensively.

Brunswik's views on this subject found only limited acceptance
on the part of psychologists, an interesting and notable exception
coming from the side of behavioral ecology, i.e., Barker (1963). Among
the various criticisms that have appeared of Brunswik's concept of
ecological representativeness, that of Hochberg (1966) deserves mention in particular, as it deals specifically with an issue of relevance to
much of the work in environmental perception, namely, the use of
photographs recommended by Brunswik in order to increase the
diversity of stimulus environments that could be presented to a
subject. Hochberg notes a variety of biases that creep into this
approach, some relating to the photographic method per se, others to
the selective factors operating where ready-made pictures obtained
from some predetermined pool (e.g., magazine, slide collections) are
employed.

Hochberg in fact rejected the very concept of representative
sampling, maintaining that even for the purpose of arriving at
ecologically valid information, the systematic variation of stimuli is to
be preferred, provided that stimuli are selected over the effective *range*
over which the relevant variables actually vary in the real environments encountered by the individual.

One of the pitfalls that this approach is prone to was not,
however, recognized by Hochberg. It is brought out by the case of the
complexity variable studied in conjunction with the man versus nature
differentiation in the writer's own research cited earlier in this paper.
As will be recalled, the attempt to apply the principles of systematic
design here was unsuccessful, since it did not appear possible to
locate stimuli of degrees of complexity in the nature and mixed
categories that were comparable to the high-complexity scenes of the
man-made set.

On the other hand, a strict "random sampling" approach to the
selection of stimuli would presuppose the possibility of dividing up
the total environment into identifiable small units and drawing a
statistically random sample from them, which is feasible at most for
sharply delimited areas, such as a single city or the equivalent.
Milgram's (1972) study of New Yorkers' pictorial recognition of different sections of their city provides a rare instance of the use of this
method. Even then the choice of views to be presented to the subject
from the selected locales remains arbitrary, unless views in all directions are taken.

This leaves two major alternative approaches. One is the adoption of a method of stratified sampling equivalent to that used in survey research for the drawing of public-opinion samples, for instance. Thus, if geographers were able to provide us with a useful catalog of different types of terrains, or features, both natural and man-made, along with information as to their distribution over geographic space, an attempt could be made to construct samples of locations representing stratified samples of the total environment. The variables involved in such stratification would have a largely descriptive status, much as Barker and associates' work on behavioral differences associated with different behavior settings. For a more functional analysis, the environmental variables would need to be supplemented by an assessment of the chosen sites along stimulus dimensions that can be meaningfully linked to the measures of aesthetic judgment to be obtained.

The second approach, which is particularly to be recommended where the investigator can specify in advance the major environmental variables of interest, is to select a limited number of locales or areas, chosen perhaps primarily for convenience. The variables of interest are then studied as they operate within each locale or area.

Note that both of the methods just proposed entail ex-post facto assessment of the sites included in the sample with respect to whatever environmental variables might be of interest to the investigator, and an analysis of their respective contributions by resort to appropriate multivariate techniques. This poses few problems in principle, except to the extent that the variables involved may in fact operate in interaction with one another; such interactions are not readily handled by the additive models on which techniques such as partial or multiple regression are based.

MODE OF PRESENTATION OF ENVIRONMENTAL STIMULI

This aspect of the methodology of research in environmental aesthetics has received more extensive discussion than the preceding, particularly with respect to the approach to the simulation of environments (e.g., photographically), and the comparability of the data thus obtained to those obtained under actual field conditions (cf. Shafer and Richards, 1974; Zube *et al.*, 1974). Suffice it to draw attention to two particular problems in this area. One relates to the limitation of photographic slides as devoid of both movement and sound. Indeed, it seems possible that the preference of natural over man-made scenes that has been encountered so frequently in this area is attributable to the greater dependence of the man-made world on sonic and dynamic components of environmental stimulation, as compared to the natural; i.e., in comparison with the latter, still photographs of man-made scenes are rather impoverished. This may likewise be one of the main reasons for the differential responses that have been found between responses to slides and to actual scenes. But what is needed is a more systematic analysis of the dimensions of these differences, e.g., by systematically introducing sound and movement, both separately and in combination. Surprisingly little use has been made of cinematography in this area of research; similarly, there is a dearth of information on the role of sonic stimulation, Southworth's (1969) study being a notable exception, though its nonsystematic design leaves the interpretation of his findings rather moot.

A second issue concerns the criteria for determining the acceptability of a given set of simulated environments. Where slides or similar stimuli are compared to the real environment, the investigator has a choice between a correlational index, and one of the differences between means for the response variables obtained in the field and under simulation conditions. The correlational approach runs into the question of how high the correlation should be; surely more than just statistical significance is demanded in this type of situation. Since generally different subjects will be tested under the two conditions, furthermore, the correlation will be obtained across environmental sites or settings, so that the results will depend strongly on the way in which these are sampled, and the range of responses obtained with respect to them under either condition. On the other hand, the differences between mean judgments can be tested for significance separately for each stimulus, but when the values are verbal ratings this approach places a heavy burden on the assumption that the two sets of judgments are comparable, in terms of the frames of reference under which the two groups were operating in making the ratings (cf. below). Even so, one suspects that this procedure, though rarely employed, will yield the most useful results for testing the effectiveness of simulation.

SELECTION OF RESPONSE MEASURES

The choice of response measures in research on environmental aesthetics has been marked by a singular lack of inventiveness and originality, a feature that has done much to hamper progress in this area of research. The very large majority of studies in this field have resorted exclusively to the use of verbal rating scales, typically without much if any attempt being made to anchor the scales for the individual, so as to ensure some degree of independence of the resulting values from the particular frame of reference or adaptation level produced by prior as well as concurrent stimulation. Nor have efforts been made to ascertain the many response biases—e.g., central tendency effects, failure to use end categories, tendency to respond in round numbers, halo effects from one judgment to another, etc.; cf. Guilford, 1954—to which these kinds of judgments are prone.

It might seem that if aesthetic responses are to be studied at all, measures of this kind, which attempt to elicit direct expression of the person's subjective evaluation of a stimulus along some verbal dimension, are necessarily called for. On further consideration, however, it is not difficult to conceive of a variety of alternative approaches that can be used instead of, or at least alongside these verbal-rating methods. Three types will be considered here: physiological, verbal, and behavioral.

Physiological Measures

Berlyne (1972a) has for some time argued for the use of certain autonomic measures in the study of aesthetics, in line with his emphasis on the concept of arousal in his theory of aesthetic response. While the amount of research that has made use of such measures is not voluminous, interesting results have been obtained with indices of arousal derived from the EEG for instance (cf. Gale, Bramley, Lucas, and Cristie, 1972; Christie, Delafield, Lucas, Winwood, and Gale, 1972). In this work the investigators have been able to demonstrate differential effects between visual and auditory complexity, and between complexity effects based on number as opposed to diversity of elements, with interesting implications for arousal-based interpretations of the role of complexity. Thus far there are virtually no instances of the application of such techniques in the realm of environmental aesthetics. An exception to this rule is represented by Wenger and Videbeck's (1968) study of pupillary response toward pictures of forest scenes, the main finding being that campers as a group show greater amounts of constriction of the pupil to forest scenes, compared to noncampers, a finding that went opposite to prediction, based on Hess's work showing pupil *dilation* in response to emotionally arousing stimuli.

Forest scenes would not, of course, be expected to elicit particularly high degrees of autonomic arousal; in fact, it would appear that stimuli employed in work on aesthetics, and environmental aesthetics in particular, would generally be characterized by relatively moderate levels of arousal.* Thus, it may be that measures of autonomic arousal may not turn out to be the most useful ones in this field, though the above-mentioned work of Gale *et al.* and Christie *et al.* suggests that judicious use of measures of cortical arousal via the EEG may well bear fruit.

A rather different kind of measure that should prove of value in work with pictures or slides are records of eye movements. Their use in visual aesthetics dates back at least four decades, as shown in Buswell's (1935) fascinating work on patterns of eye movements in response to pictorial art. They have been applied much more recently in the study of environmental perception from an information-processing view by Mackworth and Morandi (1967), with promising results, providing rather compelling evidence on the tendency of the individual to concentrate his attention on the most information-laden portions of visual displays.

Verbal Measures

Some of the limitations of the standard rating-scale measures that are the stock-in-trade of the environmental researcher have already been noted. These far from exhaust, however, the range of available types of instruments for the study of aesthetic response to the environment. Such judgments may take a variety of forms, many of which appear to be preferable, from the standpoint of psychometrics, to the rating-scale approach. They include measures of preferential choice, by the paired-comparison method or some variant of the same (e.g., method of triads), ranking methods, direct ratio-scaling methods (Stevens, 1966), and similar measures derived from psychophysics, and other scaling methods (cf. Torgerson, 1958), which not only avoid some of the biases inherent in the rating scale approach, but yield values in whose equal-interval character we can have greater confidence.

Mention might also be made of judgments derived from utility theory, involving, for instance, asking the subject to indicate how much he would be willing to pay for particular views if he had to rent or buy a house, or choose accommodations on vacation. Admittedly, these measures, as is true equally of preferential-choice measures, do

* It appears significant, if more than slightly ironic in this connection, that investigators studying the relationship between sexual arousal and aggression, (e.g., Baron and Bell, 1973) have found it expedient to use slides of the physical environment, including views of such scenic attractions as the Grand Canyon, as *neutral* control stimuli assumed to be lacking in potential for affective arousal!

not allow one to differentiate between different components of the individual's affective or evaluative response to a set of stimuli; in compensation they may prove their worth in research, analogous to that of Shafer (1969), Peterson (1967), and others, in predicting preference responses from combinations of stimulus parameters.

Behavioral Measures

Finally, a broad class of response measures should be considered that involves diverse forms of overt behavior that can be taken as direct or indirect expressions of affective response, based on interest, preference, approach-avoidance behavior, etc. These range all the way from laboratory measures of amount of voluntary exploratory activity exhibited by a subject viewing, feeling, or listening to a particular stimulus (typically in the form of data on exploration time), to indices of behavior aggregated over groups, such as number of visitors to particular locales, or number of persons stopping to view a given exhibit in a museum, or amount of time they spend doing so (cf. Melton, 1972). Many of these measures, along with more refined instrumentation such as Bechtel's (1967) hodometer, which is sensitive to pressure exerted on a given spot from the person or persons standing on it, and can thus be used to chart spatial distribution of visitors to a museum or the like, have the virtue of being unobtrusive (Webb, Campbell, Schwartz, and Sechrest, 1966). That is, they are not subject to the influences of the experimental situation, or the investigator, or the enforced judgment itself, on the individual—not an inconsequential advantage in many situations of interest to the environmental investigator. Like most overt-behavioral measures, as well as some judgmental ones as noted above, they fail to differentiate between the various components of the aesthetic response, and should probably be employed to supplement and perhaps validate, rather than supplant verbal measures. Indeed, the results of research on various aspects of aesthetics and stimulus-induced motivation make plain that an accurate picture of the individual's response to a given stimulus or situation requires the use of different sets of measures in parallel, to allow the investigator to tap different levels as well as modes of response to such stimuli.

SOME UNRESOLVED ISSUES IN ENVIRONMENTAL AESTHETICS

The field of environmental aesthetics is in such an early state, and still so lacking in systematic, theoretically guided research effort, that to talk of unresolved issues in reference to it is in a sense inappropriate. Yet, it seems worthwhile to indicate a few particular questions that are the object of controversy, and that would repay investigation in view of their theoretical as well as practical significance. Let us examine five of these.

UNITY VERSUS DIVERSITY

As Berlyne (1972a, pp. 124 ff.) has noted, the field of aesthetics has been fertile ground for the postulation of two-factor theories of beauty, involving a balance between two opposing forces or processes. Frequently, as in the writings of Leibnitz, Hegel, and Fechner, among others, the two factors have been identified as unity and variety, or in the theory of Birkhoff (1933), as order and complexity.

With the exception of an exploratory study by Berlyne and Boudewijns (1971), followed up by Berlyne (1972b), this problem has received scant attention from researchers. Berlyne *et al.* have demonstrated a tendency for pleasingness and liking to be maximal for patterns of elements that are intermediary in amount of diversity, i.e., combining similarities with respect to some features with differences in others, though Berlyne's second study, using sets of three rather than two elements, gave less clear-cut results in this regard.

The question remains, Is the principle of unity amid variety in fact reducible to such a compromise between sheer amount of similarity and difference? Or, are there more subtle structural properties that come into play, such that the effects of unity in diversity are mediated by an independent factor, i.e., that of order or patterning, based on structural relations among the elements (e.g., symmetry, regularity)? Despite the evident relevance of this issue for environmental concerns, little progress has been made even at a conceptual level in resolving it, let alone in obtaining data concerning it. Instead, we have had from different quarters an emphasis on diversity or complexity on the one hand and on unity or coherence on the other, without their being placed into relationship with each other. Nor has the semantic-differential approach, which on occasion has included scales dealing with both of these factors (e.g., Küller, 1972), provided any new insights on this question. All that this approach could tell us is whether in the individual's meaning space the factors on which these two types of scales load are the same or different; in actuality, however, the issue concerns a possible interaction between these two aspects of a stimulus configuration.

Note that this question transcends the purely theoretical level in interest, since considerations of the fittingness of a particular structure in a particular context frequently amount to questions of the extent of diversity tolerable without impairment of unity (e.g., the appropriateness of a modernistic building in a traditional neighborhood). Indeed, as we will note presently, our laws and zoning regulations make frequently implicit and occasionally explicit assumptions as to what is desirable or satisfying in this regard, which have rarely been subject to examination.

THE NATURAL VERSUS THE MAN-MADE

This issue may be differentiated into two separate ones. First, is there any inherent difference in the response to natural as opposed to man-made scenes, and if there is, what is the basis for that difference? Second, what determines the evaluation of an environment combining elements of both types as either fitting or unfitting? Both of these issues have been alluded to in the review above, the first in the context of the discussion of the role of complexity, the second under the treatment of the perception of congruity. Neither, however, has been answered very satisfactorily. With regard to the first, we saw that there appears in general to be an overall bias in favor of the natural environment, but the reasons for, as well as the generalizability of this finding remain far from clear. In regard to the second, we have done little beyond corroborating the reality of the congruity factor as such as an important one in environmental aesthetics.

Both of these issues would seem to profit from an approach aimed at specifying the stimulus correlates involved in each, i.e., the environmental–ecological attributes that differentiate the natural from the man-made world, and those that create a sense of fittingness or congruence, as opposed to their opposites, when the two are seen in juxtaposition. Undoubtedly the answers to both questions will entail a consideration of matters of content, alongside the kind of structural analysis presented. It will probably demand attention to diverse other factors as well, notably those of an attitudinal, experiential, and cultural nature, which probably go far to influence an individual's differential response to the two domains, as well as to their combination.

Just as in the case of the previous issue, it is not difficult to point to the importance of this one for applied purposes, as in the management of natural recreation areas, where questions of policy arise in regard to the establishment of man-made structures and use of man-made implements. In a similar vein the domain of the landscape architect comes to mind, with its broad array of concerns relating to the introduction of elements of nature into the realm of the man-made. Much has been written on both of these subjects, of course, but research on the individual's response is as yet largely lacking.

BEAUTIFUL VERSUS UGLY: CONTINUUM, OR N-DIMENSIONAL SPACE

While aestheticians, artists, environmentalists, and psychologists have discoursed at length on the nature of beauty, and the beautiful, little has been said about the nature of ugliness. The tacit assumption has been that, like the beast in the fairy-tale, ugliness represents the polar opposite of beauty, that is, that the two are points on a common continuum. The frequent use of semantic-differential scales in which "beautiful" and "ugly" are used to designate the end-points of the scale itself reflects this assumption.

The assumption has, however, by no means been verified, and in fact there are reasons to suspect that we may be dealing here with a multidimensional phenomenon. Ugliness, in different contexts and situations, may refer to dirt, drabness of color, lack of perceptible organization among elements, lack of proportion in size and shape (i.e., gross incongruity), etc. Indeed, there are examples from the field of the musical and visual arts, where stimulus configurations that can only be described as ugly, but placed in particular spatial or temporal contexts, are used in the service of an overall aesthetic effect that may be far from opposite to the pole of the beautiful. Note, further, the espousal of the "ugly and ordinary" by Venturi, Brown, and Izenour (1972) as an aesthetic for contemporary utilitarian architecture, as exemplified in the style of the Las Vegas strip.

Research on the stimulus correlates of ugliness might be helpful in this regard. It would most likely differentiate the determinants of this response into a number of separable components, which may at least in part be unrelated to those that a similar analysis of the correlates of beauty might uncover. This issue again transcends the realm of the theoretical in its implications. Thus, Kates (1966) among others has argued that environmental managers might do well to focus on the reduction of ugliness, rather than the cultivation of beauty, if only because consensus on ugliness is apt to be more readily achieved than on beauty. It is interesting to note that the law appears to operate from a very similar premise in its handling of legal issues touching on questions of environmental aesthetics. The individual is considered to have the right to be protected from gross insults to his senses, but not to the maintenance or creation of a positively satisfying or aesthetically appealing environment, at least in competition with property rights, etc. (Broughton, 1972).

THE EFFECTS OF PROLONGED EXPOSURE TO A GIVEN ENVIRONMENT

The large majority of the research that has been carried out in the area of environmental aesthetics has employed fairly brief presentations of particular environmental settings, typically via slides or similar simulations, and generally with observers unfamiliar with those sites. Most of these studies thus leave open the question of the changes that the individual's aesthetic responses to a particular environment undergo as a function of prolonged exposure to that environment. Yet, more so than in the case of art stimuli, for instance, this is a very central question that needs to be faced on the part of the environmental psychologist, for environments generally are experienced repeatedly, and in fact in many cases represent a virtually constant context for particular segments of an individual's life, whether around home, on the way to work and at work, or for his or her recreational activity.

The question thus arises: What is the capacity of an environment to sustain a strong positive affective response following oft-repeated exposure to it? Conversely, are strong negative responses to it similarly weakened, if not neutralized? In other words, does familiarity indeed breed contempt, i.e., a decrease in response, such as the concept of adaptation itself might suggest? We lack the answers to this question at present, but common observation suggests that in at least certain cases a particular type of environment becomes the source of increasing affect, often invested with and reinforced by symbolic significance attributed to it, as in the New Englander's affection for his hilly landscape, or the Midwesterner's no less real sense of attachment to the vast open spaces of his fields of corn or wheat (cf. Tuan, 1974, chap. 8). What limited research evidence we have, furthermore, suggests that individuals are apt to prefer types of landscapes closely resembling those to which they have become adapted; yet this is opposite to what has sometimes been assumed, on the basis of the principle that affect increases to some degree as a function of discrepancy from adaptation level (McClelland, Atkinson, Clark, and Lowell, 1953). The point is that the range over which such an increase is found is generally fairly narrow, and so we find native Eskimos actually preferring the relatively barren, arid scenes that comprise their environments to more verdent, hilly terrains (Sonnenfeld, 1967).

At the same time it appears likely that certain aspects of the environment do indeed lose their power to elicit strong arousal or affective response, particularly those based on Berlyne's collative properties. Almost by definition, novelty, incongruity, and surprise cannot be maintained for any extended period of time as the individual develops familiarity with a given environmental stimulus. To the extent that the response to complexity is indeed based on the uncertainty or conflict that a complex stimulus generates, the same principle should apply here, but it is possible that the neutralization process would progress more slowly, and perhaps not be as complete as in the case of the previous attributes. By the same token, considerations of the sustaining power of environmental stimuli render all the more cogent the emphasis given in this paper to such qualities as harmony, balance, and fittingness or congruence, rather than their opposites, since these would not necessarily change through repeated exposure.

The Place of Individual Differences in Environmental Evaluation

Most of the preceding discussion has treated the problems of environmental aesthetics without regard to factors related to the individual whose judgments of or affective responses to environmental stimuli are under consideration. Clearly, this is at most a partial approach to these problems, since here as in every aspect of behavior there are major differences among people in the way in which they respond to environments.

The question arises, How should such differences be conceptualized? In this respect it is useful to distinguish between two different approaches. On the one hand, there is the approach of the differential psychologist, or personality theorist, who might attempt to apply personality types or known dimensions of individual difference to the study of response to environmental stimuli. Applications of personal-construct theory to the realm of environmental cognition (e.g., Harrison and Sarre, 1971) represent an illustration of this approach. A somewhat different one is provided by the study of "environmental dispositions" (McKechnie, 1972; cf. also Craik, 1976), which starts out with clearly defined dimensions of individual difference that have particular reference to environmental response, in contrast to the personal-construct approach, which aims rather to uncover in an inductive fashion the dimensions used by different subjects in ordering or comparing sets of environmental stimuli.

On the other hand, it is possible to employ constructs that have specific reference to the processing of environmental stimulation, and postulate individual differences related to such processing. Two cases in point are the concepts of complexity and of adaptation level. With regard to complexity, as has been demonstrated repeatedly, there are of course wide individual differences in the levels that are preferred (Vitz, 1966; Dorfman and McKenna, 1966). Indeed, response to complexity of stimuli has been hypothesized to be a function of the individual's own position on the complexity dimension, i.e., in terms of amount of complexity that is congruent with an organism's capacity for processing information, as determined in part at least by its prior experience with stimuli varying on this dimension (cf. Dember and Earl, 1957). While the latters' view that the individual will invariably prefer a level of complexity somewhat above his own position (such that the stimulus in question can act as a "pacer") may not be correct, the point is that this kind of formulation allows us to handle individual differences in terms that are directly relatable to the stimulus dimensions of interest to the environmental aesthetician.

If one examines the work in the environmental-aesthetics area that has actually been carried out, however, one finds few studies that have made use of dimensions of individual difference that are clearly defined and integrally related to this domain of behavior. With regard to adaptation levels, one notable exception might be cited, the aforementioned study by Sonnenfeld (1967), who compared responses to diverse photographically portrayed environments varying along specified topographical dimensions of the landscape, for individuals with differing prior environmental experiences. The data from this study are unfortunately far from consistent, in part, one suspects, due to the use of preferential choice measures for pairs of photographs in which these dimensions were represented in dichotomized form.

For the most part, however, what evidence there is on individual differences in evaluative and affective responses to environments comes from investigations in which different groups of subjects, differing along such demographic dimensions as sex, age, and occupation, have been compared. Furthermore, even these studies present little information on actual differences in the evaluation of environments among the various groups, since the data are generally presented in terms of correlation matrices or factor analyses, so that the main concern is with differences among the groups in such intercorrelations, rather than in the means of the responses. The ambitious studies of Küller (1972), Lowenthal and Riel (1972a), and Zube *et al.* (1974) all fall into this category. The result is that we still lack information on systematic differences in environmental evaluation that can be related to individual differences along environmentally relevant dimensions of personality, cognitive style, attitude, and the like.

PRACTICAL APPLICATIONS OF ENVIRONMENTAL AESTHETICS

In recent years there has been a trend in the field of architecture and environmental design to downgrade aesthetics as a basis for design. This trend is readily defensible as a reaction against the one-sided emphasis of architects of the past on appearance (and frequently external appearance, to the exclusion of internal), at the expense of consideration of factors relating to use and user satisfaction and to social and community concerns. Yet, it would be erroneous to dismiss the person's affective response to the looks (and sounds and smells) of his environment as in any sense of secondary order of importance. In a purely pragmatic sense, the fact is that many of the most heated controversies in environmental management, in community governance, and in environmental legislation at all levels, concern problems of an aesthetic nature. The review, *Aesthetics in Environmental Planning*, commissioned by the U.S. Environmental Protection Agency (1973) provides further testimony, if any were needed, of the place of aesthetic factors in applied concerns.

Architecture, Design, and Planning

The relevance of aesthetics to these areas hardly needs discussion or documentation. The views of architects and planners in regard to matters aesthetic have been referred to at various points (cf. the papers by Rapoport and Kantor, 1967; Rapoport and Hawkes, 1970; and Lozano, 1974). A further paper of interest in this connection is that of Basch (1972); it is rather more critical of diverse approaches and philosophies of aesthetics applied to environmental design and planning. Basch argues for a more analytic stance in this regard, and cites (though only in a footnote) the two Rapoport articles under this heading, while focusing his discussion mainly on the work of Lynch, which he views rather critically. But it is not at all apparent how the latter's work, notably as illustrated in the *Image of the City* (Lynch, 1960), conforms to the criteria Basch proposes for the analytic approach. Yet, Basch's discussion is of value in reinforcing the place of such an approach, which one presumes may be opposed to the purely descriptive approach so prevalent in this realm, for the needs of the environmental designer and planner.

Lastly, it is worth noting the use made of aesthetics, i.e., external appearance of buildings, in the work of Newman (1973), aimed at increasing the viability of public housing projects by enhancing the sense of identity of its residents with their project. One of the ways of achieving this end that Newman has found partially successful in at least one such project involves the exploitation of diversity and distinctiveness of appearance by having the residents choose the color of their unit, and repainting the units accordingly, so as to arrive at a vastly increased sense of diversity in the appearance of the project, as well as a facilitation of the residents' identification with their particular units. A comparison of the renovated Clason Point Gardens project of Newman's with the ill-fated Pruitt-Igoe, with its unrelieved, monolithic drabness (strengthened in its latter days by the wasteland appearance conveyed by the many boarded-up apartments, broken windows. etc., along with the complete absence of landscaping) suggests that to some degree at least attention to such aesthetic considerations can indeed play a vital role in the work of the designer, even in an area such as that of public housing for the poor.

MANAGEMENT OF NATURAL RECREATION AREAS

Here, too, the relevance of environmental aesthetics should be fairly obvious, since it has repeatedly been shown (e.g., Shafer and Mietz, 1969) that aesthetic factors loom large in the motivation of those in search for a natural recreation experience, and of wilderness in particular. There are a variety of questions touching on matters of aesthetics that are closely related to some of the issues considered above that should be of direct relevance to the managers of natural recreation areas. The latter would presumably benefit from a better understanding of the basis for the power of a natural area to evoke aesthetic satisfaction, i.e., of the requirements for such satisfaction. This question is particularly pertinent to an issue that frequently arises in the natural recreation field, that of the substitutability of one environment for another that may be experiencing overcrowding or overuse.

A related question concerns the response to elements of the man-made world in the context of a natural recreation area, such as buildings (from tents or cabins to resort lodges and shops), roads (from interstate-type highways to logging or fire roads), vehicles (from buses and campers to motorboats and skimobiles, ski lifts), etc. Many

of the most hard-fought controversies in the realm of natural-recrea-
tion management relate to questions of the suitability of the intrusion
of man-made facilities, structures, or implements of this kind into a
natural area, so that questions of congruity or incongruity between the
natural and the man-made world become of primary importance.

ENVIRONMENTAL IMPACT STATEMENTS

Aesthetic considerations are mentioned as one of the three types
of factors in terms of which environmental impact of proposed
governmental projects is to be judged (the other two being environmental quality and social effects). One suspects that the relative
infrequency with which they are in fact prominently attended to is
based less on deliberate neglect of this factor than to the difficulty of
foreseeing the actual aesthetic impact of a project while it is in the
planning stage, particularly given the lack of agreement on the
determinants of aesthetic quality. Suffice it to call attention to one
specific instance in which a major issue developed over the aesthetic
effects of a proposed development on a site operated by the U.S. Parks
Service: the construction of an observation tower on the site of the
Gettysburg National Military Park. While the project was eventually
approved, it is apparent that the major reason for the controversy
surrounding it, and the resulting delay it experienced, was basically
over the effects of the erection of such a prominent landmark on a
basically flat terrain. The fact, however, is that, outside of the general
work on incongruity— which probably provides at best a very partial
account of this particular problem—we have little information on just
what the basis is for the affect elicited by such structures.

ENVIRONMENTAL LAW

Some of the most fascinating problems in applied environmental
aesthetics surface in legal discussions of the problem, in the context of
cases involving zoning laws, nuisance suits, as well as reviews of
proposed new projects of diverse kinds (cf. Bockrath, 1974; Broughton,
1972; Leighty, 1971). Two issues in particular emerge as salient in the
"legal mind's" consideration of aesthetic questions: First, there appears to be agreement that while an individual has no inherent right
to beauty, i.e., positive aesthetic quality in the environment, he does
have a right to be protected from stimuli concerning which there is
general consensus as to their character as insults to the senses,
whether in the form of noise, smells, or mere visual pollution (e.g.,

automobile junkyards; clothes drying in public view, etc.). Not surprisingly, issues involving eyesores of this latter kind have proved rather more controversial, and courts have been reluctant to grant injunctions or award damages on such grounds. On the other hand, they have increasingly looked with favor upon the use of the police power, through zoning ordinances, laws regulating billboards, etc., designed to control such eyesores, though generally on the basis of alleged hazards to health, or protection of property values, than on aesthetic grounds per se.

At the same time, courts have frequently decreed that fittingness, or congruence, among the structures built in a particular area is a valid criterion in the formulation of zoning laws, as well as in the adjucation of nuisance cases and other suits involving matters of environmental aesthetics. The point is brought out in the following two quotes from legal decisions related to billboards and junkyards respectively:

> There are areas in which aesthetics and economics coalesce, areas in which a discordant sight is as hard an economic fact as an annoying odor or sound. We refer not to some sensitive or exquisite preference but to concepts of congruity held so widely that they are inseparable from the enjoyment and hence the value of property (quoted in Leighty, 1971, p. 1394).

> Of course, equity should not be aroused to action merely on the basis of the fastidiousness of taste of complainants. Equity should act only where there is presented a situation which is offensive to the view of average persons of the community. And, even where there is a situation which the average person would deem offensive to the sight, such fact alone will not justify interference by a court of equity. The surroundings must be considered. Unsightly things are not to be banned solely on that account. Many of them are necessary in carrying on the proper activities of organized society. But such things should be properly placed, and not so located as to be unduly offensive to neighbors or to the public (quoted in Broughton, 1972, p. 462).

Here we surely have a clear link between the theoretically based experimental work in aesthetics, and an issue of the most direct practical consequence. It would seem that environmental psychologists may have something to contribute to the efforts in the legal profession to place this as yet somewhat nebulous area on a sounder footing. At the same time there is much that is of relevance for their own concerns that they may profitably study on the basis of such cases, to the benefit of their conceptualizations of these basic issues of environmental aesthetics. The same argument as to the two-way street from the basic to the applied and back again holds just as true for the other domains cited above of the practical application of problems in this area. One hopes thus for more effective interaction between the behavioral scientist and the professional and practitioner, to bring about much-needed clarifications of the issues involved. Fortunately, the problem appears to be of sufficient inherent interest to scientist and professional alike that there is good cause for optimism for the future in this regard.

CONCLUSION

"Beauty is in the eyes of the beholder." This old saw appears to be reflected in a great deal of the environmental-perception literature that has focused on individual differences in aesthetic response, or exploited such assumed differences to generate factors of aesthetic judgment from semantic-differential ratings. The present paper is dedicated to the contrary theme, that problems in the realm of environmental aesthetics can profitably be approached from the stand-point of the relationship between particular stimulus parameters and judgmental and other aesthetically relevant responses. Clearly, an integration of the two foci will be essential for fully adequate treat-ment of this complex problem area. It is worth noting, however, that without effective attention to the environmental correlates of aesthetic judgments the value of research in this area for the environmental designer or manager will be limited, indeed. For the latter (unless he is in the position of designing custom-made environments for specific customers) must necessarily keep some modal population in mind; more to the point, he will want to know in the first instance about the role played by the variables that are in fact at least partially under his control, namely, the environmental ones. It is for this reason that such stress has been placed on Berlyne's collative variables and to his theory of aesthetic response. While this theory may require modifica-tion and extension in certain respects, it represents surely the most solid base from which further developments in this area can take off.

ACKNOWLEDGMENT

This paper is based in part on a presentation by the author given at a Symposium on Experimental Aesthetics and Environmental Psy-chology, held at the International Congress of Applied Psychology in Montreal, July 1974.

REFERENCES

Barker, R. On the nature of the environment. *Journal of Social Issues*, 1963, *19*, 17–38.

Baron, R., and Bell, P. A. Effects of heightened sexual arousal on physical aggression. *Proceedings of the 81st Annual Convention of the American Psychological Association, Montreal, Canada*, 1973, *8*, 171–172.

Basch, D. The uses of aesthetics in planning: A critical review. *Journal of Aesthetic Education, 6*, 1972, 39–55.

Bechtel, R. B. *Footsteps as a measure of human preference.* Topeka, Kansas: Environmental Research Foundation, 1967.

Berlyne, E. D. The influence of complexity and novelty in visual figures on orienting responses. *Journal of Experimental Psychology*, 1958, *55*, 289–296.

Berlyne, D. E. *Conflict, arousal and curiosity.* New York: McGraw-Hill, 1960.

Berlyne, D. E. Complexity and incongruity variables as determinants of exploratory choice and evaluative ratings. *Canadian Journal of Psychology*, 1963, *17*, 274–290.

Berlyne, D. E. Arousal and reinforcement. *Nebraska Symposium on Motivation*, 1967, *15*, 1–110.

Berlyne, D. E. *Aesthetics and psychobiology.* New York: Appleton-Century, 1972a.

Berlyne, D. E. Uniformity in variety: Extension to three-element visual patterns and to non-verbal measures. *Canadian Journal of Psychology*, 1972b, *26*, 277–291.

Berlyne, D. E. (Ed.). *Studies in the new experimental aesthetics: Steps toward an objective psychology of aesthetic appreciation.* New York: Halsted Press, 1974.

Berlyne, D. E., and Boudewijns, W. J. Hedonic effects of uniformity in variety. *Canadian Journal of Psychology*, 1971, *25*, 195–206.

Birkhoff, G. D. *Aesthetic measure.* Cambridge, Mass.: Harvard University Press, 1933.

Bockrath, J. Aesthetics and condemnation awards: Problems in preserving the aesthetic environment through eminent domain. *Natural Resources Journal*, 1974, *7*, 621–633.

Broughton, R. Aesthetics and environmental law: decisions and values. *Land and Water Law Review*, 1972, *7*, 451–500.

Brunswik, E. Distal focussing of perception: Size constancy in a representative sample of situations. *Psychological Monographs*, 1944, *56* (1, Whole #254).

Brunswik, E. *Perception and the representative design of psychological experiments.* Berkeley, Calif.: University of California Press, 1956.

Buswell, G. T. *How people look at pictures.* Chicago: University of Chicago Press, 1935.

Calvin, J. S., Dearinger, J. A., and Curtin, M. E. An attempt at assessing preferences for natural landscapes. *Environment and Behavior*, 1972, *4*, 447–470.

Canter, D. An intergroup comparison of connotative dimensions in architecture. *Environment and Behavior*, 1969, *1*, 27–48.

Christie, B., Delafield, G., Lucas, B., Winwood, M., and Gale, A. Stimulus complexity and the EEG: Differential effects of the number and the variety of display elements. *Canadian Journal of Psychology*, 1972 *26*, 155–170.

Craik, K. H. Appraising the objectivity of landscape dimensions. In J. V. Krutilla (Ed.), *Natural environments.* Baltimore: Johns Hopkins Press, 1972, 292–346.

Craik, K. H. The personality research paradigm in environmental psychology. In S. Wapner, S. Cohen, and B. Kaplan (Eds.), *Experiencing the environment.* New York: Plenum, 1976, 55–79.

Crozier, J. B. Verbal and exploratory responses to sound sequences varying in uncertainty level. In D. E. Berlyne (Ed.), *Studies in the new experimental aesthetics.* New York: Halsted Press, 1974, 27–90.

Cullen, G. *Townscape.* London: Architectural Press, 1961.

Day, H. Evaluations of subjective complexity, pleasingness and interestingness for a series of random polygons varying in complexity. *Perception and Psychophysics*, 1967, *2*, 281–286.

Dember, W. N., and Earl, R. W. Analysis of exploratory, manipulatory and curiosity behaviors. *Psychological Review*, 1957, *64*, 91–93.

Dorfman, D. D., and McKenna, H. Pattern preferences as a function of pattern uncertainty. *Canadian Journal of Psychology*, 1966, *20*, 143–153.

Downs, R. The cognitive structure of an urban shopping center. *Environment and Behavior*, 1970, *2*, 13–39.

Ertel, S. Exploratory choice and verbal judgment. In D. E. Berlyne and K. B. Madsen (Eds.), *Pleasure, reward, preference*. New York: Academic Press, 1973, 115–132.

Fabos, J. G., Careaga, R., Greene, C., and Williston, S. *Model for landscape resource assessment. Part I of the "Metropolitan Landscape Planning Model"* (METLAND). Amherst, Mass.: University of Massachusetts, Department of Landscape Architecture and Regional Planning, 1973.

Fiske, D. W., and Maddi, S. A conceptual framework. In D. W. Fiske and S. Maddi (Eds.), *Functions of varied experience*. Homewood, Ill.: Dorsey Press, 1961, 11 56.

Gale, A., Bramley, P., Lucas, B., and Christie, B. Differential effect of visual and auditory complexity on the EEG: Negative hedonic value as a crucial variable: *Psychonomic Science*, *17*, 1972, 21–24.

Governor's Conference on Natural Beauty. *The Governor's Conference on Natural Beauty*, September 12, 13, 1966. Community Center: Hershey, Pa., 1966.

Guilford, J. P. *Psychometric methods*. New York: McGraw-Hill, 1954.

Harrison, J., and Sarre, P. Personal construct theory in the measurement of environmental images: Problems and methods. *Environment and Behavior*, 1971, *3*, 351–374.

Hendrix, W. G., and Fabos, J. G. Visual land use compatibility as a significant contributor to visual resource quality. *International Journal of Environmental Studies*, 1975, *8*, 21–28.

Hochberg, J. Representative sampling and the purposes of research: Pictures of the world, and the world of pictures. In K. Hammond (Ed.), *The psychology of Egon Brunswik*. New York: Holt, Rinehart & Winston, 1966, 361–381.

Iltis, H. H. A plea for man and nature. (Letter). *Science*, 1967, *156*, 581.

Iltis, H. H. The optimum human environment and its relation to modern agricultural preoccupations. *The Biologist*, 1968, *50*, 114–125.

Jacobs, J. *The death and life of great American cities*. New York: Vintage, 1961.

Jacobs, P., and Way, D. How much development can landscape absorb? *Landscape Architecture*, 1969, *59*, 296–298.

Kaplan, R. Predictors of environmental preference: Designers and "clients." In W. F. E. Preiser (Ed.), *Environmental design research*. Vol. 1. Stroudsburg, Pa.: Dowden, Hutchinson & Ross, 1973, 265–274.

Kaplan, R. Some methods and strategies in the prediction of preference. In E. H. Zube, G. G. Fabos, and R. O. Brush (Eds.), *Landscape assessment: Values, perceptions and resources*. New York: Halsted Press, 1975.

Kaplan, S. An informal model for the prediction of preference. In E. H. Zube, G. G. Fabos, and R. O. Brush (Eds.), *Landscape assessment: Values, perceptions and resources*. New York: Halsted Press, 1975.

Kaplan, S., Kaplan, R., and Wendt, J. S. Rated preference and complexity for natural and urban visual material. *Perception & Psychophysics*, 1972, *12*, 334–356.

Kates, R. W. The pursuit of beauty in the environment. *Landscape*, Winter 1966/67, *16*, 21–25.

Koestler, A. *Insight and outlook: An inquiry into the common foundations of science, art and social ethics.* New York: Macmillan, 1949.

Küller, R. *A semantic model for describing perceived environment.* Stockholm: National Swedish Institute for Building Research, 1972.

Leighty, L. L. Aesthetics as a legal basis for environmental control. *Wayne Law Review,* 1971, *17,* 1347–1396.

Leopold, L. Landscape aesthetics. In A. Meyer (Ed.), *Encountering the environment.* New York: Van Nostrand Reinhold, 1971, 29–46.

Leopold, L. B., and Marchand, M. O'B. On the quantitative inventory of the riverscape. *Water Resources Research,* 1968, *4,* 709–717.

Litton, R. B. Jr., Aesthetic dimensions of the landscape. In J. V. Krutilla (Ed.), *Natural environments.* Baltimore: Johns Hopkins Press, 1972, 262–291.

Lowenthal, D., and Riel, M. Milieu and observer differences in environmental associations. New York: American Geographic Society, 1972a. (*Publications in Environmental Perception, 7*).

Lowenthal, D., and Riel, M. The nature of perceived and imagined environments. *Environment and Behavior,* 1972b, *4,* 189–207.

Lozano, E. Visual needs in the urban environment. *Town Planning Review,* 1974, *45,* 351–374.

Lynch, K. *The image of the city.* Cambridge, Mass.: MIT Press, 1960.

Mackworth, N. H., and Morandi, A. J. The gaze selects informative details within pictures, *Perception and Psychophysics,* 1967, *2,* 547–552.

McClelland, D. C., Atkinson, J. W., Clark, R. A., and Lowell, E. L. *The achievement motive.* New York: Appleton-Century, 1953.

McKechnie, G. E. A study of environmental life-styles. Unpublished PhD Dissertation, University of California (Berkeley), 1972.

Melton, A. W. Visitor behavior in museums: Some early research in environmental design. *Human Factors,* 1972, *14,* 393–403.

Meyer, L. B. *Emotion and meaning in music.* Chicago: University of Chicago Press, 1956.

Milgram, S. A psychological map of New York City, *American Scientist,* 1972, *60,* 181–194.

Mindus, L. The role of redundancy and complexity in the perception of tone patterns. Unpublished MA thesis, Clark Univeristy, 1968.

Nairn, I. *The American landscape.* New York: Random House, 1963.

Newman, O. *Defensible space.* New York: Macmillan, 1973.

Nunnally, J. C., Faw, T. T., and Bashford, M. B. Effect of degrees of incongruity on visual fixations in children and adults. *Journal of Experimental Psychology,* 1969, *81,* 360–364.

Peterson, G. L. A model of preference: Quantitative analysis of the perception of the visual appearance of residential neighborhoods. *Journal of Regional Science,* 1967, *7,* 19–30.

Rapoport, A., and Hawkes, R. The perception of urban complexity. *Journal of the American Institute of Planners,* 1970, *36,* 106–111.

Rapoport, A., and Kantor, R. E. Complexity and ambiguity in environmental design. *Journal of American Institute of Planners,* 1967, *33,* 210–222.

Sanoff, H. Visual attributes of the physical environment. In G. J. Coates and K. M. Moffett (Eds.), *Response to the environment.* Raleigh, N.C.: School of Design, North Carolina State University, 1969, 37–60.

Santayana, G. *The sense of beauty.* New York: Scribners, 1896.

Schwarz, H., and Werbik, H. Eine experimentelle Untersuchung über den Einfluss der syntaktischen Information der Anordnung von Baukörpern entlang einer Strasse auf Stimmungen des Betrachters. *Zeitschrift für experimentelle und angewandte Psychologie*, 1971, *18*, 499–511.

Shafer, E. L., Jr., Hamilton, J. F., Jr., and Schmidt, E. A. Natural landscape preferences: A predictive model. *Journal of Leisure Research*, 1969, *1*, 1–19.

Shafer, E. L., Jr., and Mietz, J. Aesthetic and emotional experiences rate high with northeast wilderness hikers. *Environment and Behavior*, 1969, *1*, 187–198.

Shafer, E. L., Jr., and Richards, T. A. A comparison of viewer reactions of outdoor scenes and photographs of those scenes. U.S. Department of Agriculture, Forest Service *Research Paper* NE-302, 1974.

Shephard, P. *Man in the landscape: A historic view of the aesthetics of nature*. New York: Knopf, 1967.

Sonnenfeld, J. Environmental perception and adaptation-level in the arctic. In D. Lowenthal (Ed.), *Environmental perception and behavior*. Chicago: Department of Geography, University of Chicago, 1967, 42–59. (*Research Paper #109*).

Southworth, M. The sonic environment of cities. *Environment and Behavior*, 1969, *1*, 49–70.

Steinitz, C., and Way, D. A model for evaluating the visual consequences of urbanization. In C. Steinitz, and P. Rogers (Eds.), *Qualitative values in environmental planning: A study of resource use in urbanizing watersheds*. Section III. Washington, D.C.: Office of Chief of Engineers, Department of the Army, 1969.

Stevens, S. S. A metric for the social consensus. *Science*, 1966, *151*, 530–541.

Torgerson, W. J. *Theory and methods of scaling*. New York: Wiley, 1958.

Tuan, Y.-F. *Topophilia: A study of environmental perception, attitudes and values*. Englewood Cliffs, N.J.: Prentice-Hall, 1974.

U.S. Environmental Protection Agency. *Aesthetics in environmental planning*. Washington, D.C.: Office of Research and Development, U.S. Environmental Protection Agency, 1973.

Venturi, R. *Complexity and contradiction in architecture*. New York: Museum of Modern Art, 1966.

Venturi, R., Brown, D. S., and Izenour, S. *Learning from Las Vegas*. Cambridge, Mass.: MIT Press, 1972.

Wathen-Dunn, W. (Ed.), *Models for the perception of speech and visual form*. Cambridge, Mass.: MIT Press, 1967.

Webb, E. J., Campbell, D. T., Schwartz, R. D., and Sechrest, D. *Unobtrusive measures*. Chicago: Rand McNally, 1966.

Wenger, W. D., and Widebeck, R. Pupillary response as a measure of aesthetic reaction to forest scenes. Syracuse: College of Forestry, State University of New York, 1968. (*Report* No. 1, *Project K.*)

White, M., and White, L. *The intellectual versus the city*. Cambridge, Mass.: Harvard University Press, 1962.

Wohlwill, J. F. Amount of stimulus exploration and preference as differential functions of stimulus complexity. *Perception and Psychophysics*, 1968, *4*, 307–312.

Wohlwill, J. F. Human response to levels of environmental stimulation. *Human Ecology*, 1974, *2*, 127–247.

Wohlwill, J. F. Children's responses to meaningful pictures varying in diversity: Exploration time vs. preference. *Journal of Experimental Child Psychology*, 1975, *20*, 341–351.

Wohlwill, J. F. Complexity and other stimulus determinants of preference and interest for scenes from the outdoor environment. Paper in preparation, 1976.

Zube, E. H., Pitt, D. G., and Anderson, T. W. *Perception and measurement of scenic resources in the Southern Connecticut River Valley*. Amherst, Massachusetts: Institute for Man and his Environment, University of Massachusetts, 1974.

3

Perception of Landscape and Land Use

ERVIN H. ZUBE

Air, water, fire, and earth were declared by Aristotle to be the four essential elements of life. Recent history has clearly demonstrated the significance of his declaration. As public awareness developed that all was not well with our environment, and as supporting evidence accumulated during the past two decades, attention was focused initially on the first two of these classically defined elements, air and water. Relationships between air and water quality and public health and safety were defined and, where possible, quantified and steps were taken to establish standards that were deemed to be consonant with desired and acceptable health and safety conditions. These standards, promulgated primarily through the legislative process, were based both on professional judgments and empirical data. The standards were defined in terms of physical parameters of the elements such as parts-per-million, units of biochemical oxygen demand (BOD), or units of discharge per unit of time.

Most recently, the emphasis has shifted to fire, or expressed in a broader current context, to energy. Concern with energy supply and demand appears to have taken precedence, as a matter of public policy, over air and water quality and has resulted in the relaxation of quality standards and in the extension of deadlines for meeting those standards.

ERVIN H. ZUBE · Institute for Man and Environment, University of Massachusetts, Amherst, Massachusetts.

Inextricably intertwined with the policies, laws, and management objectives for these three classical elements, however, is the fourth, the earth or the land. We have come to realize that the quality of air and the quality of the water are strongly influenced by the way in which we use the land. The geographic location of different uses (e.g., upstream, downstream, upwind, downwind, valley, ridge-top), the relationship between uses (e.g., heavy industrial–residential, conservation–solid waste disposal, conservation–recreation), and the management practices applied to the land (e.g., fertilizer application for agriculture, clear-cutting for timber production, vegetation, and top-soil stripping for subdivision development) all influence air and water quality as well as our perception of the quality of the environment. In like vein, the development of many of our energy resources has raised questions of the impact of development on the land as well as on air and water quality. The aesthetic impact of strip mining, the future economic uses of mined-over lands and standards for mined-area rehabilitation are, for example, hotly debated issues.

In contrast to the definitions for air and water quality, scientists, engineers, and policymakers have not been able to define land quality on as objective a basis or to devise standards that can be expressed in such seemingly precise terms as parts-per-million or BOD. The search for a qualitative definition for the classical element earth—the landscape—has, thus far, resulted in a more diffuse and subjectively oriented definition and a set of guidelines that are based more on professional judgments than on empirical data. These differences in qualitative definitions between air, water, and land can be attributed, at least in part, to the inherent characteristics of the elements and to the nature of the relationship between man and each of the elements. Air and water, for example, have traditionally been looked upon as common resources or "free goods." As such they could be used and abused at will as they belonged to everyone and hence to no one. They were also therefore most susceptible to the establishment of uniform national policies, including qualitative standards, to be implemented at the state and local levels. Land, however, is not a free good. All of the land is owned. Approximately two thirds of the land in the 50 states is privately owned. The remainder is in federal ownership. The right to set policy and establish controls on the use of the privately owned land is constitutionally vested in state governments, but these powers have historically been passed down to local units of government, resulting in thousands of uncoordinated policy decisions. Land has also been viewed in policy and practice as a collection of discrete parcels rather than as a resource continuum. For example, both in the marketplace and in the making of local land-use decisions, land has traditionally been treated on a tract by tract or parcel by parcel basis. The land has been viewed as a commodity.

A RESEARCH RATIONALE

Why then should one be concerned with the study of landscape and land-use perception? There are several compelling reasons ranging from the theoretical to the applied. At the most general level, landscape and land-use perception research should contribute to the building and delineation of a theory of man–environment relations. It should encompass the development of concepts and ordering schema for ideas and data relative to the macro or geographic scale, for identifying individual and group characteristics that are important in shaping the environment, for identifying the nature and magnitude of the effect of the environment on people, and for defining the mechanisms that link people and environment (Rapoport, 1973, p. 125). The macro or geographic scale referred to above ranges from the scale of neighborhoods to larger conceptual regions (Saarinen, 1969, 1974) associated with resource management and regional planning programs.

In a more applied tenor, the value and significance of such research lies in its relationship to specific planning and management programs and to specific public policies. It should attempt to provide environmental planners, designers, managers, and policymakers with a better understanding of the perceptual processes of their public constituencies. Matters of concern to these individuals include the range of perceived resource and land-use values of their several publics and the extent of agreement among these various publics on the strength and the direction of those perceived values. For example, do conservationists and prodevelopment constituencies agree on the identification of lands to be protected because of their scenic value or on the appropriateness or compatibility of proposed adjacent land uses? The significance of such data lies in the insights and understanding they provide relative to the design and implementation of landscape management and land-use policies and practices. To paraphrase Firey's hypothesis (1960), for resource policies and plans to be anything more than opportunistic ventures, they must be *perceived* by the concerned publics to be ecologically possible, ethnologically adoptable, and economically gainful.

The discussions that follow in this chapter are organized around two topics that are drawn from this research rationale. The first is that of selected public policy issues that are deemed to be of particular significance to landscape management and land-use planning and that the author considers to be important perceptual issues as well. That is, the efficacious implementation of these policies will be, at least in part, dependent upon an understanding of the "public's" perceptions of the issues. The second topic is that of practice and research. The practice to be discussed refers to planning efforts to implement specific policies. The research to be considered includes studies that have been generated to facilitate understanding of the public's perceptions of policy implications, to identify the public's values, and to test the efficacy of implementing certain practices: the congruence of professionals' proposals with the perceptions of their clients.

The following operational definitions will be used throughout the chapter. *Policy* refers to the criteria that decision makers use in deciding what to do or what not to do. *Land use* is defined as the activity that is carried on a given piece of land and that is supported by specific physical structures and management practices. *Landscape* refers to the combined physical attributes of the environment. *Scenic (or aesthetic) value* refers to the perceived visual value an individual or group places on the landscape; it is a product of an interaction between man and the landscape.

One final preliminary note on that which is to follow: the focus is on that section of the landscape that starts at the edge of metropolitan areas and encompasses the rural, forested, and wilderness countrysides. The focus is on those sectors of the landscape that are still subject to both extensive and intensive change through resource-management programs and to land-use decisions that are responsive to the pressures of urban and suburban growth.

MATTERS OF POLICY

Concern with scenic values can be traced historically through public policy, particularly in reference to the development of the National Park System, for well over one hundred years. One of the earliest examples is the cession of Yosemite Valley for use as a park to the state of California by the Congress in 1864 (Zube, 1973a). Within the past decade, however, it has also been a matter of primary concern at the federal level in the Highway Beautification Act of 1965, in the National Wild and Scenic Rivers Systems Act of 1968, and in the National Trails System Act of 1968. The highway act was addressed to the problem of ameliorating the misfit or the ugly while the rivers and trail acts were addressed to the protection and preservation of unique scenery or landscapes of special value. The passage of the Multiple Use—Sustained Yield Act of 1960 directed the Forest Service to combine economic and environmental objectives in the administration of their lands. The act presented the service with a new set of problems in "forest aesthetics" (Shafer, 1967).

Further attention has been focused on the issue of scenic or aesthetic landscape values through the provisions of the National Environmental Policy Act of 1969 (NEPA), the Coastal Zone Management Act of 1972 (CZM), the Principles and Standards for Planning Water and Related Land Resources (P and S) (Water Resources Council, 1973), and the several bills introduced in the Congress over the past five years to establish a national land-use policy, including the two recent introductions of 1975, the draft bill by Senator Henry Jackson (Land Resources Planning Assistance Act) and HR 3510 by Representative Morris Udall (Land Use and Resource Conservation Act).

Section 102(2)(b) of the NEPA requires that all agencies of the federal government shall "identify and develop methods and procedures. . . which will insure that presently unquantified environmental amenities and values may be given appropriate consideration in decision making along with economic and technical consideration." Section 102(2)(c) of the act requires that all agencies shall include an environmental impact assessment "in every recommendation or report or proposals for legislation and other major Federal actions significantly affecting the quality of the human environment. . . ." The CZM act encourages states to identify, along with other objectives, critical areas within the coastal zone and to establish procedures for the preservation or restoration of these areas because of their conservation, recreation, ecological, or aesthetic values. The P and S set forth a multiple objective approach to comprehensive studies and regional or river basin planning of water and related land resources involving federal participation that is consonant with the mandates of the NEPA. Primary objectives defined in the P and S are (1) national economic development and (2) environmental quality. A specified component of the latter objective is the "Management, protection, enhancement or creation of areas of natural beauty and human enjoyment such as open and green space, wild and scenic rivers, lakes, beaches, shores, mountain and wilderness areas and estuaries. . . ."

The net effect of these legislative and executive actions has been to thrust environmental values into a planning and decision making process that heretofore had been primarily based on economic values and technological considerations. These actions have also forced the consideration of land as a natural resource continuum as well as a commodity. The concept of critical environmental areas, including scenic areas, as a key component in land-use planning and the notion of an ecological orientation to land-use planning have gained momentum at both the federal and state levels.

The concept of critical environmental areas is one that has been a key component of most proposed national land-use policy legislation. The bill passed by the Senate in the 93rd Congress (U.S. Senate Bill 268) defined critical areas as

> lands where uncontrolled or incompatible development could result in damage to the environment, life or property, or the long term public interest which is of more than local significance.

A number of states have already initiated management or identification programs for critical areas, including California, Colorado, Florida, Illinois, Indiana, Minnesota, Missouri, North Carolina, Oregon, Virginia, and Wisconsin (Smithsonian, 1974, p. 251). A region-wide inventory has also been conducted for the six New England states (New England Natural Resources Center, 1974). Factors or variables common to many of these programs are: wildlife habitats, natural/scientific/ecological areas, scientific/geological areas, wilderness recreation areas, agricultural land, historical areas, scenic areas, flood areas, fire and earthquake areas, and areas predisposed toward air pollution (Smithsonian, 1974, pp. 52–58). Thus, the critical areas concept that is of central interest to many states and the CZM program also encompasses the concerns of the scenic rivers and trails legislation, and of many aspects of the environmental quality objective of the P and S. The primary emphasis is on the identification of critical areas and their subsequent preservation or protection from undesirable effects of land-use change and development.

Both the NEPA and the P and S address the process of planning as well as the planning product. Environmental quality planning objectives and impact statement requirements have stimulated alternative planning equations and procedures that set the environment as the independent variable, in contrast to traditional equations and procedures in which population and related economic growth is the independent variable. (McHarg, 1969; Lovejoy, 1973; Greenwood and Edwards, 1973; Twiss, 1973; Bailey, 1974). Efforts are made to define the ecological suitability of the land for proposed uses and activities, to identify the carrying capacity of the landscape resource, or the intensity of use that can be accommodated without diminution of existing resource qualities, and to analyze alternative spatial land-use relationships so as to identify those specific relationships that minimize adverse impacts and enhance beneficial impacts. In other words, the concept of the land as a resource as well as a commodity begins to acquire an operational definition under the procedures followed in this alternative planning equation.

PRACTICE AND RESEARCH

Within recent years, a number of state of the art reviews on landscape perception of scenic values have emphasized application or practice (Fabos, 1971; Zube, 1973a; Stanford Research Institute, 1973; Dunn, 1974; Cerny, 1974). These papers focus on assumed and empirically based scenic values and on methods for integrating these values into planning practice and impact analysis. A review by Arthur, Daniel, and Boster (1975), while also focused on aesthetic values, went one step further and probed the question of the relationship between economic and aesthetic values. And, finally, a volume by Zube, Brush, and Fabos (1975) presents recent developments in research, planning, and management within the context of a broad range of values—economic, historic, conservation, recreation, and scenic that influence people's perceptions of the landscape.

All of these reviews have several characteristics in common: (1) They emphasize scenic or aesthetic values of the landscape; (2) they tend to deal with application more than with theory; and (3) they respond, directly or indirectly, to a common set of public policies, promulgated primarily during the past decade, that relate to the issue of scenic values.

The following discussion of practice and research is organized in two sections: (1) landscape management issues in public forests, especially in the American West, and with an emphasis on the aesthetic consequences of alternative management practices; and (2) regional resource and land-use planning especially in the northeastern United States.

LANDSCAPE AND LAND-USE ISSUES IN THE FOREST SERVICE

The Multiple Use—Sustained Yield Act of 1960 provided the policy directive for the Forest Service to institute land management practices that recognized both environmental and economic values, including timber production, livestock grazing, wildlife, aesthetics, and recreation. The act provides in Section 4(a) that management decisions be made

> . . .with consideration being given to the relative values of the various resources and not necessarily the combination of uses that will give the greatest dollar return or the greatest unit output.

The nature of the management issues related to aesthetic values, for example, are clearly illustrated in the public controversy surrounding the land-management policies on the Bitterroot National Forest in the late 1960s (U.S.D.A. Forest Service, 1970). In 1968 members of the Recreation Subcommittee of the Ravalli County, Montana, Resource Conservation and Development Committee expressed their very great concern about forest-management practices on the Bitterroot to the Forest Service and to Senator Lee Metcalf of Montana. They were concerned about: (1) clear-cutting, which they felt accelerated soil erosion and created long-lasting eyesores; (2) terraces cut into the hillsides for machine planting of trees, which allegedly results in low seedling survival, threatened soil stability, lowered productivity of the land, and aesthetically displeasing conditions; (3) the building of too many poorly located and improperly constructed roads, causing sedimentation and reduced natural beauty; and (4) the overemphasis on timber production, at the expense of aesthetic values and watershed protection (U.S.D.A. Forest Service, 1970, p. 1; U.S. Senate, 1970, pp. 15–16). Such were the set of environmental and aesthetic problems presented to the Forest Service. It is of interest to note the consistency with which physical environmental concerns such as sedimentation, soil stability, and watershed protection are linked with suggested visual correlates of natural beauty and aesthetic or scenic value. A primary issue identified by the Select Study Committee (U.S. Senate, 1970) was the professionals' and nonprofessionals' divergent perceptions of the nature of the problems, as exemplified by judgments of compatible and appropriate uses of the forest (U.S. Senate, 1970, p. 14).

Controversies such as this have stimulated two distinct but related, albeit modest, attempts to come to grips with the intentions of the multiple-use management policy. First is the use of multidisciplinary teams for the drafting of management plans, including landscape architects who are charged with responsibility for the aesthetic component.* Such multidisciplinary teams are intended to represent the

---

* The U.S.D.A. Forest Service is now the largest single employer of landscape architects in the United States.

broader range of values and judgments of appropriate uses intended in multiple-use management. Guidelines and manuals are also prepared, based on the values and judgments of the professionals, for management approaches to aesthetic problems (U.S.D.A. Forest Service, 1968, 1973, 1974; Litton 1974a, 1974b); and for the adoption of tools and techniques particularly oriented to the identification and definition of scenic resources (Amidon and Elsner, 1968; Litton, 1968; Elsner, 1971; Litton, 1973).

Second, there has been an attempt to identify the public's perceived aesthetic values and to test professionals' assumptions and judgments, thereby providing a base of empirical data in support of aesthetic management objectives. Craik (1969, 1972a), for example, tested the interobserver objectivity of a system of landscape dimensions developed by Litton (1968, 1972) for the identification of scenic resources in national forests. A graphic landscape typology and a series of rating scales were developed based on Litton's landscape dimensions. The rating scales included viewer position, extent of view, foreground-middleground-background, composition, panoramic characteristics, direction of lighting, enclosure, isolated forms, surface shape, focal characteristics, and clouds. Craik found an "impressive amount of agreement" in the use of the rating scales for describing specific scenes (1972a, p. 300). He further found high agreement on the use of the rating scales among expert and nonexpert panels. Rank order correlations between the expert and nonexpert panels were: landscape architecture and university panels: +.66; Forest Service and university panels: +.67; combined expert panels and university panels: +.72; and between the two expert panels, landscape architecture and Forest Service panels: +.83.

Shafer and colleagues (Shafer, Hamilton, and Schmidt, 1969; Shafer and Thompson, 1968; Shafer and Mietz, 1970, Shafer and Tooby, 1973) have developed a predictive regression model of landscape preferences. Utilizing Adirondack Mountain campers' preference ratings of black and white photographs of 100 scenes from all over the United States, a model was developed for identifying preferred scenic recreation areas in forested and natural landscapes. The model related preference scores to three zones of the photograph, immediate, intermediate, and distant*; and to major vegetation, nonvegetation (ground and grass), and water. In a more recent study, Brush and Shafer (1975) illustrated the application of the model to

---

* The immediate, intermediate, and distant zones used by Shafer are analogous to the Litton/Craik zones of foreground, middleground, and background.

landscape management problems that primarily involve changes in amounts of vegetation and water. The model is intended for use in predicting the scenic impact of activities such as large-scale forest-management programs involving timber-harvest and water-resources management programs relating to the creation of new water bodies and the alteration of existing water bodies.

Brush has also initiated a number of studies on the relationships of forests and forest-related values to midwestern and eastern urban and metropolitan areas, including an investigation of the influence of forest vegetation on perceived distance of objects (Brush, 1974) and a study of the relationship of silvicultural practices to the visual enhancement of forests for recreational purposes (Brush, 1975). Drawing upon the design skills of the landscape architect and implicitly suggesting the need for an interdisciplinary approach, he developed a set of guidelines that relate management techniques (e.g. clear-cutting, even-aged management, uneven-aged management, and single tree or group tree selection for harvesting) to visual qualities that are adjudged to enhance the recreationist's environmental experience.

Daniel and Boster (Boster and Daniel, 1972; Daniel, Wheeler, Boster, and Best, 1973; Daniel and Boster, 1974), have investigated the perception of a range of timber management or treatment procedures, using an approach based on the Theory of Signal Detection (TSD). Measurement techniques based on TSD permit evaluation of differential preferences and differing criterion states of individuals or groups. In reference to the perception of timber-management treatments the issues are: (1) Do observers make differential aesthetic responses to the various treatments; and (2) can observers discriminate reliably among various treatments, e.g., clear-cut, strip-cut, conventional logging, unlogged, etc. (Daniel et al., 1973)? Boster and Daniel (1972) found clear empirical evidence that perception of the various treatments of ponderosa pine forests in the Southwest "varies markedly" over treatments and that "there is at least a general tendency for the more natural appearing areas to be preferred. . . ." In subsequent studies Daniel and Boster (1974) investigated variations in perceptions of 26 user, interest, and professional groups ($N=680$) to six areas representing a cross-section of management options in Arizona's ponderosa pine national forests. The groups distinguished between the six management options and the author's report: "Perhaps the most striking feature. . .is the extent to which there was agreement in the scenic preferences of these diverse interests" (p. 39).

Daniel (1974) also reported on the relationship of scenic quality to landscape features that are susceptible to management manipulation.

From a list of 27 features two primary factors were identified using step-wise multiple regression analysis and factor analysis. A general "downed wood" factor (e.g., logging residues) consisted of variables that have large negative effects on scenic quality. A second "tree density factor" consisted of variables such as tree density, average tree diameter, and variability of tree density. These findings suggest that cleanup practices after logging operations affect perceived scenic quality in ponderosa pine forests, and that management and cutting practices that enhance variability in tree stand composition and distribution would be preferred to management and cutting practices that support uniform even-aged stands.

The Multiple Use–Sustained Yield Act has apparently had an impact on forest-management practices. It has stimulated a modest, but important array of perception-oriented research aimed at providing managers and policymakers with a better understanding of the perceptual processes of their publics and at providing insights into the values of various forest-using publics; information that can contribute to the modification and implementation of management policies. The requirements of the NEPA have also provided an added stimulus for understanding the aesthetic impact of alternative forest-management practices on different kinds of forests in different regions.

A LANDSCAPE PLANNING ORIENTATION

A shift in geographic and policy focus from the Rocky Mountain region and from forest management to the northeastern United States and regional landscape planning programs provides a setting for a discussion of another set of policy-related studies. The studies had their genesis in several resource-planning programs ranging in scale from 50 square miles to 167,000 square miles: from the island of Nantucket (IN) (Zube and Carlozzi, 1966) and the United States Virgin Islands (USVI) (Zube, 1968) to the southeastern New England region (SENE) (Zube and Fabos, 1972; Riotte, Fabos, and Zube, 1975) and the North Atlantic Region (NAR) (Zube, 1970).

Each of these studies involved the identification of critical areas and each also involved the identification of scenic values within a broad resource-planning context. While several of the studies were antecedent to the formal definition and statement of the policies discussed earlier, they are indicative of planning trends of the mid-to-late 1960s that helped to set the stage for those policies.

Critical areas, sometimes referred to as major biophysical resources (IN) or special values resources (USVI), included unique or rare biological communities, scenic areas, significant historical and archaeological sites, wildlife and marine habitats, fragile ecosystems—those particularly sensitive to human activities—and environmental process areas such as flood plains and aquifer recharge areas. Scenic values in each instance were identified and, in the NAR and SENE studies, ranked on the basis of professional or expert judgments.

The NAR and SENE studies were also multiple-objective studies and employed the kind of approach to planning set forth in the P and S. Both studies addressed the traditional federal resource-planning objective of economic development and both also addressed the new objective of environmental quality.

Substantive involvement in studies such as these confront both the practitioner and the researcher with an interesting and challenging array of issues and questions relating to: expert and nonexpert values, planning methods and procedures, the efficacy of implementing related public policies, and the relationship of application to theory (see Zube, Pitt, and Anderson, 1975, pp. 151–152; Zube, Brush, and Fabos, 1975, pp. 59–64). A number of these issues have been addressed in the studies outlined in Table 1. Before discussing some of the findings, however, an overall research structure should be identified and several important connecting links between the resource-planning programs and the research studies as well as between the individual research studies should be noted.

Table 2 shows the relationships between the individual research studies in reference to four organizing schemata that provide a general structure and continuity to the research. Three of these schemata—intergroup agreement, assessment of planning procedures, and the prediction of scenic quality—can also be used as an organizing framework for the previously discussed landscape and land-use issues in the Forest Service. A number of studies, notably those by Craik and Daniels and Boster, relate to the theme of intergroup agreement. Those same studies, together with those of Shafer and Brush, also assess planning and management procedures of foresters, landscape architects, and recreation planners. And, finally, the work of Daniels and Boster and of Shafer and Brush is directed toward the development of predictive models for scenic quality and for management practices and procedures that enhance visual amenities.

The northeastern United States, southern Connecticut River valley, Lorne coastline, and crossdisciplinary studies each have as a primary objective, an assessment of intergroup agreement on the description and/or evaluation of the landscape. Each study also includes, as participants, at least one group who represents a profession

TABLE 1

SUMMARY OF RESEARCH STUDIES RELATED TO RURAL-REGIONAL LANDSCAPE PLANNING

Study	Primary objectives	Participants	Display media	Response formats
A. Northeastern U.S. landscapes (Zube, 1973b)	1. To assess the extent of agreement among groups on the description and evaluation of landscapes of the northeastern U.S. 2. To test planners' assumptions on physical indicators of scenic quality	1. Resource managers/planners ($N = 19$) 2. Resource managers/planning students ($N = 24$) 3. Environmental designers ($N = 50$) 4. Environmental design students ($N = 20$) 5. Environmental technicians ($N = 19$) 6. Secretaries ($N = 30$) 7. Others ($N = 23$) (TOTAL $N = 185$)	1. 35-mm color slides of landscapes ($N = 18$) 2. 35-mm color slides of landscape drawings ($N = 9$)	1. Paired comparison of slides 2. Semantic scales
B. Southern Connecticut River valley landscapes (Zube, Pitt, and Anderson, 1974, 1975)	1. To determine the extent of agreement among groups on the evaluation and description of different landscapes, using both field experience and photographic representations 2. To identify physical landscape dimensions hypothesized to be determinants of scenic	1. Resource managers/planners, 2 groups ($N = 50$) 2. Environmental designers, 2 groups ($N = 45$) 3. Professional engineers ($N = 33$). 4. Office and clerical workers, 2 groups ($N = 55$) 5. Undergraduate design students ($N = 23$)	1. Wide-angle color photographs ($N = 56$) 2. Panoramic color photographs ($N = 8$) 3. Site visits ($N = 8$)	1. Q-sort of wide-angle photographs 2. Rank ordering of panoramic photographs 3. Semantic scales 4. Landscape feature checklist 5. Personal history questionnaire

Table 1. *continued*

Study	Primary objectives	Participants	Display media	Response formats
	resource values and which can be measured using aerial photography and topographic maps 3. To analyze the relationships between subjects' evaluative responses and quantified physical dimensions	6. Undergraduate psychology students (N = 18) 7. Suburban-rural residents, 2 groups (N = 48) 8. High school students (N = 17) 9. Center-city residents (N = 11) (TOTAL N = 301)		
C. Granby, Massachusetts landscape (Cooper, 1974)	1. To provide a means for residents of a community to assess and map the scenic quality of their town	1. Local residents (N = 40)	1. Wide-angle color photographs (N = 31)	1. Q-sort wide-angle photographs
D. Lorne coastline, Australia (Zube and Mills, 1975)	1. To assess crosscultural agreement on the evaluation of coastal landscapes of Australia	1. Residents of the Lorne coastal area (N = 25) 2. Australian visitors to the Lorne coastal area (N = 76) 3. University of Massachusetts graduate students (N = 22) (TOTAL N = 123)	1. Panoramic color photographs (N = 24)	1. Q-sort panoramic photographs 2. Personal history questionnaire

Study	Objectives	Subjects	Displays	Methods
E. Crossdisciplinary understanding and communication in resource planning (Zube, 1974a)	1. To assess the efficacy of using remote data sources for landscape resource studies 2. To assess interprofessional agreement on the use of descriptive and evaluative terms for the landscape	1. Resource managers (N = 30) 2. Environmental designers (N = 60) (TOTAL N = 90)	1. Aerial photographs (N = 6) 2. Topographic maps 3. Site visits	1. Semantic scales 2. Free description 3. Rank ordering of aerial photographs 4. Personal history questionnaire
F. Predicting scenic values from physical landscape dimensions (Anderson, Zube, and MacConnell, 1975)	1. To develop a predictive model for scenic quality utilizing the findings from Study B and an expanded sample of the landscape	1. Graduate students in landscape architecture (N = 30)	1. Wide-angle color photographs (N = 217) 2. Aerial photographs 3. Topographic maps	1. Q-Sort of wide-angle photographs
G. Landscape classification (Palmer and Zube, 1975)	1. To retest environmental designers scenic values using displays from Study B 2. To identify perceived landscape/land-use similarities	1. Hampton Institute, Va., architecture students (N = 10) 2. Graduate students in landscape architecture (N = 10) 3. Undergraduate psychology students (N = 18) (TOTAL N = 38)	1. Wide-angle color photographs (N = 56)	1. Q-Sort of wide-angle photographs for: a. scenic quality b. perceived similarity 2. Free description

TABLE 2
RELATIONSHIPS BETWEEN STUDIES

Study	Intergroup agreement	Assessment of planning procedures	Prediction of scenic quality	Replication of display media
A. Northeastern U.S. landscapes	X	X	X	
B. Southern Connecticut River valley landscapes	X		X	X
C. Granby, Massachusetts, landscape		X		X
D. Lorne coastline, Australia	X			
E. Crossdisciplinary understanding and communication in resource planning	X	X		
F. Predicting scenic values from physical landscape dimensions			X	X
G. Landscape classification		X		X

involved in multidisciplinary resource-planning teams (e.g. environmental designers, landscape architects, resource managers, engineers). Thus, each study also addresses questions of the extent of agreement among experts and nonexperts or among experts representing different professions. The Lorne study provides some insight into crosscultural landscape perception.

The northeastern United States, Granby, crossdisciplinary, and landscape classification studies also have as a primary objective the assessment of planning concepts or procedures. Indirectly, they begin to address the question of the efficacy of scenic resource policy implementation from an operational planning point of view. The display media used in the northeastern study were selected from or were based on photographic reference materials collected during the conduct of the NAR study. The assumptions that were tested as to the significance of physical landscape dimensions, land form and land-use pattern, as indicators of scenic quality were also from the NAR. The crossdisciplinary study was designed to test the efficacy of the data-collection procedures used in the conduct of the NAR, and to assess the degree of communication and understanding about the landscape among several of the professions represented on the planning team. The landscape classification study was intended to probe the perceptual validity of a basic concept of the IN, NAR, and SENE studies, the idea that there are identifiable visual attributes or characteristics associated with different land uses and vegetative-cover patterns and that these characteristics provide a basis for classifying the landscape for planning purposes. The Granby study applied techniques developed in the southern Connecticut River valley study to a specific town. The study was developed so as to provide a means for local citizens to map the scenic values of their town as one component of a natural resources inventory.

The interest in developing predictive capabilities for scenic quality stems from the initial NAR physical-dimension assumption and is one that has considerable significance in terms of identifying critical areas for alternative management programs such as those designated in the IN and USVI. The Connecticut River valley and predictive studies both represent more systematic and quantitative attempts than the initial NAR effort.

The final organizing schema, or in this case more appropriately a unifying thread, is a basic set of landscape displays consisting of 56 wide-angle color photographs that have been used in four of the seven studies. All 56 displays were used in the southern Connecticut River valley, the predictive, and the landscape classification studies. In the latter study they were included in a larger set of 217 displays. In Granby, a sample of 10 of the displays were included with a set of local photos. In each study, however, the use of the basic set, in total or in part, provides an important basis for making interstudy comparisons and inferences.

INTERGROUP AGREEMENT

Do the environmental design experts on planning teams represent an aesthetic elite when applying their professional judgments to landscape description and evaluation, or are their values consonant with those of other interested publics? Fines (1968) and Laurie (1975) suggest that while experts do not differ from others in their value orientations, they are, however, by virtue of their training and experience, able to draw finer distinctions and perceive a wider range of values.

Environmental design experts were included in each of the four studies included under this theme. In general, the findings are similar to those of Craik (1972) and Daniel and Boster (1974). They tend to support a hypothesis of similar evaluative judgments. There are, however, some interesting variations that also merit mention. In the study of northeastern U.S. landscapes, product–moment correlations on scenic evaluation of the 27 displays for the 7 participant groups ranged from .42 to .86 ($r = .49, p < .01$). Of the 21 correlations, 6 were below .70 and ranged from .43 to .68. In the Connecticut River valley study, correlations on scenic evaluation for the 13 participant groups on the rank ordering of the 8 paroramic displays ranged from .47 to 1.00 ($r = .76, p < .01$). As many as 85% of the correlations were greater than .90; correlations on the Q-sort of the 56 wide-angle displays ranged from .22 to .96 ($r = .22, p < .01$); and 74% of the correlations were greater than .80.

In the northeastern U.S. study, product–moment correlations on the use of semantic scales for describing 4 landscape displays ranged from .58 to .96 ($r = .62, p < .01$). However, 20 of the 21 correlations were above .72. In the Connecticut River valley study, correlations on the use of descriptive scales (see Table 3) for 8 landscapes ranged from .59 to .97 ($r = .56, p < .01$);85% of the correlations were above .82. Semantic scales were also used in the crossdisciplinary study involving two participant groups of environmental designers (one participating in field studies and one in office studies of the same five sections of landscape using maps and aerial photographs) and one group of resource managers and planners (office studies).* A high degree of consensus is indicated by correlations that ranged from .89 to .91 ($r = .49, p < .01$). Modest professional orientations or biases were discernible, however, from an analysis of written reports. Each subject was asked to describe one section of the landscape as if "you were writing a letter to a friend who has never seen the region." The designers referred to vistas and views and spatial composition characteristics more frequently, landscape attributes that are obvious concerns of their profession.

The resource managers and planners displayed a similar professional bias and referred to watershed characteristics, roads, forestry, and recreation opportunities more frequently than their designer counterparts. Both professions tended to refer to land-use activities and patterns such as towns, open space, and farms and to topography with nearly identical frequency.

In January of 1975 a planning program was initiated for the shoreline of Lorne, Australia, including a study of scenic quality using the techniques and procedures developed in the Connecticut River valley study (Mills, 1975). The panoramic photographs of the Lorne shoreline were used in a subsequent crosscultural study with graduate

---

* The list of semantic scales used in this and subsequent studies was developed from a pool of 3,800 descriptive objectives elicited from 85 subjects responses to 18 35-mm slides representative of northeastern U.S. landscapes.

TABLE 3
SEMANTIC DIFFERENTIAL SCALES

Simple—Complex	A		E
Beautiful—Ugly	A	B	E
Diverse vegetative cover—Uniform vegetative cover			E
Bright—Dull	A	B	E
Varied—Monotonous	A	B	E
Inviting—Uninviting		B	E
Hard—Soft	A	B	E
Flat—Mountainous	A		E
Urban—Rural	A		E
Orderly—Chaotic			E
Distant—Intimate	A		E
High scenic value—Low scenic value		B	E
Smooth—Rough	A	B	E
Natural—Man-made		B	E
Colorless—Colorful	A	B	E
Great—Small			E
Closed—Open		B	E
Angular—Rounded		B	E
Artificial—Natural	A		E
Unity—Variety			E
Obvious—Mysterious	A	B	E
Dynamic—Static			E
Wet—Dry	A		E
Pleasant—Unpleasant	A	B	E
Like—Dislike		B	E
Common—Unusual		B	
Tidy—Untidy		B	
Boring—Interesting		B	
Light—Dark		B	

A: Indicates used in study A, northeastern U.S. landscapes.
B: Indicates used in study B, southern Connecticut River valley landscape.
E: Indicates used in study E, crossdisciplinary understanding and communication.

students in landscape architecture (Zube and Mills, 1975). Correlational analysis of the Q-sort data for the three participant groups suggests similar patterns of perceived scenic quality by nonprofessional Australians and aspiring American professionals (Table 4).

General patterns of association among groups were strikingly similar in each of these studies. There is, overall, a generally impressive indication of agreement on scenic qualities of the landscape among participant groups. While experts or professionals did not generally exhibit discernible differences in descriptive or evaluative responses from most other groups, there are a few notable exceptions.

TABLE 4
LORNE STUDY: BETWEEN GROUP CORRELATIONS

	Landscape architecture graduate students		Lorne year-round residents	
Landscape architecture graduate students	—	—		
Lorne year-round residents	.76*	.72**	—	—
Lorne seasonal and nonresidents	.87*	.83**	.89*	.90**

* Pearson product–moment correlation, $N = 24$; $r = .50$; $p < .01$.
** Spearman rank-order correlation, $N = 24$; $r_s = .508$; $p < .01$.

In the northeastern U.S. study, two groups accounted for the six lowest correlations on scenic evaluation ($r = 43$ to .68), the secretarial and the environmental technician groups. Four of the lowest correlations were with the environmental designer and the environmental-design student groups.

One of the two secretarial-office worker groups in the Connecticut River valley study accounted for eight of the 20 correlations that were less than .80. They were, however, all within the .7 and .8 range and at best minimally divergent from the norm of other groups. More pointed are the low correlations (below .5) with every other group of a small group ($N = 11$) of black center-city residents of the Q-sort scenic evaluation. Of the 12 correlations, 6 were .30 or lower.

As inconclusive as these findings are, they do suggest a note of caution, and that agreement among experts and nonexperts on landscape resource evaluation may be mediated by socioeconomic and cultural factors. They suggest that the prudent decision maker would be well advised to carefully consider the perceptual processes of the publics to be served by a proposed planning program, particularly if those publics include diverse ethnic, cultural, and socioeconomic groups.

PLANNING PROCEDURES

A second objective of the northeastern U.S. study was to test the efficacy of assumptions made in the NAR as to the use of land form, land-use pattern, and water features as predictive dimensions of scenic quality. Scale values for these dimensions for each display were provided by a panel of five judges. The relationships of perceived scenic values of the seven participant groups with these dimensions tended to support the NAR assumptions when applied to more natural landscapes, but not when applied to more man-made (suburban) landscapes. The findings also indicated the desirability of defining more objective, quantitative scales for the dimensions and the necessity of searching out a broader array of dimensions that might relate to the more man-made sectors of the landscape on the edges of metropolitan areas. Correlations for the seven participant groups with the predicted values of the 27 displays on the basis of the combined scale values for the three dimensions ranged from .05 to .33 ($r = .49$, $p < .01$). When only the 13 most natural displays were included, however, the correlations for the seven groups ranged from .52 to .71 ($r = .68, p < .01$).

A second objective of the crossdisciplinary study was to investigate the efficacy of using aerial photographs and topographic maps as primary data sources for regional landscape studies, a procedure that was extensively followed in the NAR and to a lesser degree in the IN, USVI, and SENE studies. Each of the three participant groups of professionals (two environmental design and one resource management) completed an identical set of tasks, including: (1) describing each of four sections of the landscape using a set of 25 semantic scales; (2) describing a section of the landscape in a "free" written statement; and (3) rank-ordering a set of aerial photographs of the study area on the basis of scenic quality.

One group of environmental designers undertook these tasks in the field as though they were conducting a regional landscape survey. The other two groups were provided with aerial photographs and topographic maps in their offices. The findings, as discussed earlier, provided strong support for the use of such remote data for large area studies. They also indicated that field analysts tend to concentrate more on the "view from road" than on generalizing to the broader landscape. An important operational implication of these findings is that the major resource inventory data sources, topographic maps and aerial photographs, used by other professionals on the planning team also have considerable utility for the study of the landscape, and that it is probably not necessary to "directly see" all of the landscape to make reasonable inferences for planning purposes. It is important, however, as with any inventory procedure employing remote data sources, to obtain ground truth.

The Granby study was designed to test the Q-sort technique used in the Connecticut River valley study on an applied resource inventory project being conducted by the U.S. Soil Conservation Service in the town of Granby, Massachusetts. The objective of such resource inventories is to provide local decision makers with quantitative and qualitative information on the town's natural resource base. These data are to be used by the local planning board, conservation commission, and board of selectmen in making more informed and environmentally sound land-use recommendations (Zube and Isgur, 1975).

A sample of 21 color, wide-angle (35 mm) photographs, selected so as to represent major landscape differences within the community were combined with 10 photographs from the Connecticut River valley study for the Q-sort deck. The 21 Granby photos were intended to serve as marker scenes to which other sectors of the town would be compared in order to develop a generalized map of scenic quality. The 10 photos from the earlier study were intended to provide a basis for comparing the evaluative responses of the Granby residents with those of the broader Connecticut River valley study population. The findings suggest a very strong parallel between the two groups with a rank-order correlation of .94 ($p < .01$) and a product–moment correlation of .98 ($p < .01$) for the 10 displays.

The 31 photographs were rank-ordered by mean value and divided into four groups for purposes of serving as marker scenes. Visual analysis of each photograph identified dominant landscape characteristics and features related to the four qualitative groupings. The highest-ranked group was dominated by open water and topography in natural settings. The next highest group included more diverse mixes of vegetation including wetlands, open fields defined by wood lots, the traditional New England townscape with village green, and less pronounced topography than the previous group. The next to the lowest group showed additional signs of man-made elements and in general quite flat topography. The man-made elements consisted of houses arrayed along straight subdivision streets and the addition of utility lines to both village greens and more rural settings. The lowest-ranked group consisted of strip commercial developments, a gravel borrow pit, and major intrusions of transmission lines.

The results of this qualitative content analysis reinforce those of a similar exercise involving the 56 displays from the Connecticut River valley study wherein it was concluded:

> . . .cultural artifacts and land management practices are important determinants of scenic resource values: they are an indication of the care and conscious planning and design that must accompany the introduction of any man/made feature into the landscape, particularly if the feature is one which will be perceived as out of context or in sharp contrast with the existing landscape. (Zube *et al.*, 1974c, p. 12.)

The final study to be reported on under the schema of planning procedures is one that addresses a common planning practice, one that was followed in the NAR and SENE studies, the practice of land-use or landscape classification. The reasons for classification are obvious: it provides a means of organizing information and for simplifying the complex. In these regards, a landscape classification system should have considerable potential value for decision makers, particularly if it is one that is also understood and accepted by his constituent public (Coleman, 1973, p. 4). Broad land-use classifications are found, for example, in the Hawaii land-use planning program (Eckbo, 1973) and in land-use proposals for California (Heller, 1971). The Hawaii land-use plan classifies all lands under one of four categories: urban, rural, agricultural, and conservation. The proposal for California also has all lands contained under four categories: agricultural, conservation, urban, and regional reserve. The classification schema developed for the NAR and SENE studies were oriented to descriptive, visual landscape characteristics, and included categories related to forests, towns, farms, and urban areas. An assumption behind the use of the NAR and SENE schema was that such categories would evoke regionally generalized visual images and thus facilitate communications between planners and publics, particularly in reference to alternative management strategies under consideration for the different landscape classes (see Cox, Haught, Zube, 1972; Zube, 1974b). Whether or not such visual images are evoked, how individuals would aggregate or classify landscapes, and the extent of agreement on such classifications are questions that speak to the efficacy of such schema for certain uses in planning programs.

There is some evidence to support this notion that certain landscapes do evoke images or at the least are associated with identifiable physical and social characteristics. These images or characteristics are also of considerable importance to planning and management programs and to perceptions of the quality of the environment. The wilderness landscape presents such an example. The idea or image of wilderness carries with it certain expectations of environment and experience for the user (Lucas, 1964; Stanky, 1972; Veal, 1973). Stanky has suggested that effective management of wilderness landscape requires the identification of both an ecological carrying capacity and a sociological carrying capacity (1972, p. 99) and that the sociological carrying capacity is influenced by use levels, type of use, spatial-temporal variations, user behavior, and perception of resource quality. Chambers (1974) found that the perceived quality of a residential/

recreational environment was affected by perceived population density—or, more precisely, by perceived density of structures—and that change in the quality of subdivided land is at least partly density-dependent.

These studies suggest the possibility that there may be thresholds that help to define or classify certain kinds of landscapes; thresholds expressed in changes in numbers of people, numbers of structures, amount of tree cover, kinds of activities, and/or land-use patterns; and that these thresholds are related to both the perceived quality of that environment and the image or perceptual classification (e.g., rural or suburban) in which one places it.

An initial step in the direction of developing a perceptually based landscape classification schema was undertaken using the displays from the Connecticut River valley study (Palmer and Zube, 1975). Three panels of judges were asked to: first, Q-sort the 56 displays according to scenic quality, replicating one of the tasks from the earlier study; second, to sort the same 56 displays according to their similarity; and, third, to provide descriptive comments about what was most representative of each group of similar landscapes. Correlational analysis of the scenic quality Q-sort means for the three panels of judges with the three most similar subgroups from the Connecticut River valley study provided very nearly mirror-image coefficients of the earlier findings with the majority of the coefficients of the earlier findings with the majority of the coefficients at the .80 level or higher. Collapsing all subgroups from both studies into single composite populations yielded a scenic quality correlation of .97.

Data from the perceived similarity Q-sort were analyzed by means of a paired association matrix and cluster analysis. The matrix indicates, for all subjects, the frequency with which each scene is paired with every other scene. The cluster analysis using the paired association matrix data then groups the scenes according to the distance between them in terms of factor scores. Seven groups or clusters were identified as follows:

1. *Farms:* rural lands under management, including rural housing, farm buildings, agricultural fields, and pastures.

2. *Meadows and woods:* natural-appearing open areas, including abandoned fields, natural meadows, and surrounding woods.

3. *Forested hills:* landscapes dominated by dense woods, including obvious topographic variation.

4. *Open water:* landscapes dominated by large areas of open water, including reservoirs, lakes, and rivers.

5. *Wetlands and streams:* natural-appearing landscapes that include visible wetlands and small streams or rivers.

6. *Towns:* intensive-use areas, including suburban residential developments, town centers, and commercial areas.

7. *Industry:* areas influenced by the presence of heavy industrial developments.

These initial findings tend to support the concept of a landscape continuum ranging from the most natural (e.g., forested hills, wetlands, and streams) through various stages of management (e.g., meadows and farms) to the more nearly totally man-made landscape (e.g., industry and towns). They also tend to partially support the way in which land use has traditionally been classified for planning purposes but suggest that in some instances a dominant resource such as topography may be as important as the dominant use for classification purposes.

PREDICTING SCENIC QUALITY

In addition to the development of management procedures for specific sites as previously suggested by the work of Daniels and Boster (1974), Shafer, Hamilton and Schmidt (1969), and Brush and Shafer (1975), another primary objective for the development of predictive capabilities for scenic quality is to enable the planner or manager to identify the range of scenic values over the surface of his area of responsibility (e.g., Fines, 1968; Linton, 1968; Craik, 1972b; Wright, 1973). One of the purposes of such broad-scale assessments is to provide a basis for comparing scenic resource values and patterns of distribution with the relative values and distribution patterns of other resources such as soils, minerals, aquifers, and forest vegetation. Such comparisons provide a basis for identifying critical resource areas, areas with concentrations of resource values.

The initial efforts in the northeastern U.S. study to assess the effectiveness of physical dimensions as indicators of scenic quality led to a more quantitative and systematic approach in the Connecticut River valley study. As Table 5 shows, six categories comprising 23 individual dimensions were identified and measurements were obtained for each dimension for each of the 56 displays. Multiple regression analysis using mean scenic values as the dependent variable and the physical dimensions as independent variables was used to determine which dimensions best explained or predicted scenic values (Table 6).

TABLE 5
LANDSCAPE DIMENSIONS

Land form

Relative relief ratio—the range of vertical elevations (based on sample points) per unit area.
Absolute relative relief—the range of vertical elevations (based on sample points) within the view area.
Mean slope distribution—the mean of a random sample of slopes, the steepness of land form.
Topographic texture—the degree of dissection of land surface, the drainage density.
Ruggedness number—the roughness of land form based on absolute relative relief, mean slope, and topographic texture.
Spatial definition index—the amount of enclosure created by landform.
Mean elevation—the mean of a random sample of elevations.

Land-use area

Land-use diversity—the relative areal distribution of land uses within the view.
Naturalism index—the degree of naturalism as indicated by land use.
Percentage tree cover—the amount of land covered by trees per unit area.

Land-use edge

Land-use edge density—the amount of edge created by adjacent land uses per unit area.
Land-use edge variety—the number of edge types per view.
Land-use compatibility—an indication of the visual congruence of adjacent land uses.

Land-use contrast

Height contrast—the difference in height of the dominant elements of adjacent land uses.
Grain contrast—the difference in the size of individual elements of adjacent land uses.
Spacing contrast—the difference in the spatial distribution of the elements of adjacent land uses.
Evenness contrast—the difference in height, size, and distribution of elements of adjacent land uses.
Naturalism contrast—a measure of the difference in naturalism of adjacent land uses.

Water

Water edge density—the amount of land/water edge per unit area.
Percentage water area—the amount of surface water per unit area.

View

Area of view—the size of the view area.
Length of view—the maximum length of view.
Viewer position—the relative vertical position of the viewer to the view.

TABLE 6
PREDICTIVE COMPARISONS

Dimension	Conn. R. study N = 56	Unstratified N = 217	Naturalism stratification Less than 3 N = 30	3 to 6 N = 71	More than 6 N = 96
Absolute relative relief			1.1		3.1
Mean slope distribution	9.8				
Topographic texture	1.2		1.3		
Ruggedness number		2.7			10.7
Spatial definition index	4.1				
Mean elevation		1.5		11.3	
Land-use diversity					2.1
Naturalism index	3.1	37.5	16.4	11.0	16.2
Percentage tree cover	1.6	1.3	1.1		1.4
Land-use edge density	2.2				3.2
Land-use edge variety	1.6		2.1		
Land-use compatibility	23.6	3.9	1.5	1.7	
Height contrast	6.7				
Grain contrast	4.1		7.7	4.1	3.1
Spacing contrast	2.0		2.2		
Evenness contrast	1.9				
Naturalism contrast	1.3				
Water edge density		3.1			12.1
*Water area density				1.9	
Area of view			24.4	1.0	
Length of view		1.7		5.9	3.0
Viewer position					2.7

* Water area density was substituted for percent of water area in the Connecticut River study.
Note: Values are shown only for those dimensions that contributed 1.0% or more to the explanation.

The 56 displays represented a limited sample of the landscape, however, particularly in reference to settings containing water, to areas on the metropolitan fringe, to transitional land-use areas, and suburban settings. The predictive scenic value study, including the original 56 displays, was designed to increase the landscape sample substantially ($N = 217$) and to test the reliability of the scenic values of the original landscape sample (Anderson, Zube, MacConnell, 1975). The increased landscape sample size also allowed for stratification of the displays so as to investigate the variable relationships of dimensions to natural landscapes and to more man-made landscapes.

The scenic values of the large sample were established, for purposes of the regression analysis, using the Q-sort procedure, by a panel of 30 graduate students in landscape architecture. Correlation analysis provided a reliability test, comparing mean scenic values for the total population ($N = 307$) for the 56 original displays with the mean values from the panel of 30 for the same displays ($r = .92$).

The results of the several regression analyses are indicated in Table 6 and are compared with that from the Connecticut River study. The results of both an unstratified regression, using all 217 displays, and three stratified regressions are shown. The stratified regressions are based on the notion of a landscape continuum ranging from natural to man-made. The degree of naturalism is indicated by the naturalism index and includes: (1) those settings that are most natural as indicated by a naturalism index of less than 3 ($N = 50$); (2) those settings that are most man-made as indicated by a naturalism index of greater than 6 ($N = 96$); and (3) those that comprise the middle ground between 3 and 6 ($N = 71$). It must be kept in mind, however, that the most man-made displays in this context are those that occur in suburbia and on the metropolitan fringe. They are not representative of the center city.

Of interest in these several regressions are the shifts that occur as a function of the larger sampling of the landscape and as a result of the stratifications.* The two most dramatic shifts are in the relative strengths of land-use compatibility and naturalism index as predictors. Naturalism appears to be the strongest predictor in each case and land-use compatibility becomes a very modest predictor at best. In all but the most natural regression one or more of the land-form dimensions also account for a major portion of the explained variability. In the most natural regression, the area of view is the strongest predictor bespeaking the importance of vistas, panoramas, and view points in the landscape.

In each regression at least seven dimensions explained 1% or more of the total variance. The top seven dimensions for each regression accounted for the following percentages of the total variation:

Connecticut River valley study —54.6%
Unstratified large sample   —51.7%
Most natural (e.g., 3)   —55.6%

* Changes were made in the measurement techniques for 16 of the dimensions after the Connecticut River valley study. Ten of the changes have no significant effect on dimension values (correlations of old and new values were .84 or higher). The changes may, however, have contributed to the reduced predictive strength of spatial definition index, height contrast, evenness contrast, and naturalism contrast suggested for the larger sample in Table 6.

Middle ground (e.g., 3–6)   —37.9%
Most man-made (e.g., 6)   —51.5%

These data suggest that more difficult areas for predicting scenic quality using physical dimensions are those that fall between the more natural and the more man-made, those areas where perhaps the dynamics of land-use change are most evident. These data also suggest that further stratifications along the lines indicated in the landscape classification study may produce greater understanding of the relationships of physical landscape attributes and characteristics with perceived quality.

FUTURE DIRECTIONS

Directions for future research have been alluded to thus far in reference to: (1) identifying potential relationships of ethnic, cultural, and socioeconomic variables with environmental perception; (2) identifying physical correlates of perceived landscape quality, particularly in reference to the more man-made or urban landscape; (3) investigating potential relationships between perceived density of population or structures, landscape image, and perceived quality; and (4) developing perceptually based landscape classification schemata that have utility for land-use planning activities. Additional research issues and needs include: (1) exploration of the effectiveness of various simulation techniques (e.g., photographs, movies, drawings, and models) and surrogates for "real-world" experience; (2) investigation of the perceived compatibility of different land uses and their relationships to landscape quality; (3) work in the replication of findings to address the question of reliability, including both longitudinal and cross-sectional studies of both landscapes and persons; and (4) develop a vocabulary and feedback system for making research findings comprehensible and usable by environmental policymakers, planners, designers, and managers.

SUMMARY

A number of compelling reasons for studying landscape and land-use perception were discussed in the introduction to this chapter, including:

1. To develop concepts and ordering schemas for ideas and data relative to the macro or geographic scale.

2. To identify individual and group characteristics that are important in shaping the environment.
3. To identify the nature and the magnitude of the effect of the environment on people.
4. To define the mechanisms that link people and environment.
5. To provide designers, managers, and policymakers with an understanding of the perceptual processes of their public constituencies.

A number of current and important landscape and land-use planning issues were also identified, issues that are related explicitly or implicitly to laws and legislation, judicial decrees, regulations and executive orders, or professional standards; to those actions and instruments that constitute public policy at the federal and state levels, such as:

1. The inclusion of "environmental amenities" or "visual and cultural values" in regional resource-planning programs and in environmental impact studies.
2. The definition of critical environmental areas, including critical scenic areas.
3. The concern with ecological suitability or the carrying capacity of the land for proposed uses and activities.

The research that was subsequently discussed is related to many, but by no means all of these reasons and issues in both direct and indirect ways. From a theoretical viewpoint, this body of research begins to suggest some group characteristics as they relate to specific management and planning issues in general geographic areas. It also begins to suggest some of the landscape dimensions and characteristics that affect people's perceptions of different environments.

From an applied viewpoint, it indicates several potential systematic approaches to the identification of critical scenic areas or critical components of the landscape to be considered in land management and planning programs. It also represents approaches based on publics' values, and hence hopefully begins to indicate to decision makers that data can be generated that have heretofore been considered to be so subjective and idiosyncratic as to be both unobtainable and unquantifiable. To the extent that such data can be demonstrated to be neither arbitrary nor capricious, they should in the future also have potential value in support of legal actions relating to cases of land-use conflict. This is an important point of consideration as it represents the ultimate authority in land-use decision making. In the final analysis disputes and conflicts related to perceived incompatible land uses are resolved in the courts.

The endeavor to relate physical dimensions to perceptual processes recognizes the physical resources with which the land manager or planner works, and attempts to establish a link between perceptions of quality of the resource, and the way in which the resource is manipulated and managed.

And, finally, the notion of landscape image or character as a basis for land-use classification schema presents an interesting and potentially valuable research challenge. Also, the concept of a perceived carrying capacity for the landscape as well as an ecological carrying capacity, if substantiated and applied to a wider range of landscape classes, could make an important contribution to both theory and practice: contributing to the understanding of mechanisms that link people and environment; identifying the nature and the effect of some sectors of the environment on people; and helping to define planning and management policies that are more consonant with various publics' perceptions of environmental quality.

REFERENCES

Amidon, E. L., and Elsner, G. H. *Delineating Landscape View Areas. . . A Computer Approach*. U.S.D.A. Forest Service Research Note PSW-180, Pacific Southwest Forest and Range Experiment Station, Berkeley, Calif., 1968.

Anderson, T. W., Zube, E. G., and MacConnell, W. Predicting scenic quality. In E. H. Zube (Ed.), *Studies in landscape perception*, Institute for Man and Environment, University of Massachusetts, Amherst, 1976.

Arthur, L. H., Daniel, T. C., and Boster, R. S. *Scenic beauty assessment: A literature review*. Tucson, Arizona, U.S.D.A. Forest Service Research Paper (draft), Rocky Mountain Forest and Range Experiment Station (pre-print), 1975.

Bailey, R. G. *Landscape-capability classification of the Lake Tahoe basi, California—Nevada, a guide for planning*, South Lake Tahoe, Calif., U.S.D.A., Forest Service, 1974.

Boster, R. S., and Daniel, T. C. Measuring public response to negative management. In *16th Annual Watershed Symposium Proceedings*, Phoenix, Arizona, 1972, 38–43.

Brush, R. O. *The effect of forest vegetation on perceived distance*. Unpublished manuscript, 1974.

Brush, R. O. *Spaces within the woods: Managing forests for visual enjoyment*. Unpublished manuscript, 1975.

Brush, R. O., and Shafer, E. L. Application of a landscape preference model to land management decisions. In E. H. Zube and R. O. Brush and J. Gy. Fabos (Eds.), *Landscape assessment: Values, perceptions and resources*. Stroudsburg, Pa.: Dowden, Hutchinson and Ross, Inc., 1975.

Cerny, W. Scenic analysis and assessment. *CRC Critical Reviews in Environmental Control*, June 1974, 221–250.

Chamber, A. D. Assessment of environmental quality in relation to perceived density in recreational settings. *Man-Environment Systems*, 1974, 4(6), 353–360.

Coleman, A. A new approach to the classification of landscapes. *Landscape Research News*, England, Winter, 1973–4, 1(6), 3–4.

Cooper, R. *An application and evaluation of three landscape planning models for the town of Granby, Massachusetts* (Terminal project report). Department of Landscape Architecture and Regional Planning, University of Massachusetts, Amherst, 1974.

Cox, P. T., Haught, A. L., and Zube, E. H. Visual quality constraints in regional land use changes. *Growth and Change, a Journal of Regional Development*, 1972, 3(2), 9–15.

Craik, K. H. Human responsiveness to landscape: An environmental psychological perspective. In K. Coates and K. Moffett (Eds.), *Responses to environment*. Student Publication of the School of Design, Vol. 18, Raleigh, North Carolina State University, 1969, 168–93.

Craik, K. H. Appraising the objectivity of landscape dimensions. In J. Krutilla (Ed.), *Natural Environments*. Baltimore, Md.: Johns Hopkins University Press, 1972a, 292–346.

Craik, K. Psychological factors in landscape appraisal. *Environment and Behavior*, 1972b, 4(3), 255–266.

Craik, K. H., and Zube, E. H. (Eds.). *Perceiving Environmental Quality*. New York: Plenum Publishing Co., 1976.

Daniel, T. C., Wheeler, L., Boster, R. S., and Best, P. R., Jr. Quantitative evaluation of landscapes: An application of signal detection nalaysis to forest management alternatives. *Man-Environment Systems:* 3(5), September 1973, 330–344.

Daniel, T. C., and Boster, R. S. *Measuring scenic beauty: The SBE method*, Rocky Mountain Forest and Range Experiment Station, Tucson, Arizona (pre-print), 1974.

Daniel, T. C. Psychological scaling of scenic quality of forest landscapes. Paper presented at Rocky Mountain Psychological Association meetings, Denver, Colorado, May 1974.

Dunn, M. C. Landscape evaluation techniques: An appraisal and review of the literature. Working paper No. 4, Center for Urban and Regional Studies, University of Birmingham, England, April, 1974.

Eckbo, G. Open space and land use. In Lovejoy, D. (Ed.), *Land use and landscape planning*. New York: Barnes and Noble, 1973, 233–247.

Elsner, G. H. *Computing visible areas from proposed recreation developments*. U.S.D.A. Forest Service Research Note PSW-246, Pacific Southwest Forest and Range Experiment Station, Berkeley, Calif. 1971.

Fabos, J. G. An analysis of environmental quality ranking systems. In *Recreation Symposium Proceedings, U.S.D.A. Forest Service Experiment Station, Upper Darby, Pa., 1971*, 40–55.

Fines, K. D. *Landscape evaluation: A research project in East Sussex. Regional Studies*. Vol. 2. Oxford, England: Pergamon Press, 1968, 41–55.

Firey, W. *Man, mind and land: A theory of resource use*. Glencoe, Ill.: Free Press, 1960.

Greenwood, N., and Edwards, J. M. B. *Human environments and natural ecosystems*. North Scituate, Mass.: Duxbury Press, 1973.

Heller, E. (Ed.). *The California tomorrow plan*. Los Altos, Calif.: William Kaufmann, Inc., 1971.

Laurie, I. Aesthetic factors in visual evaluation. In E. H. Zube, R. O. Brush, and J. G. Fabos (Eds.), *Landscape assessment: Values, perceptions and resources*. Stroudsburg, Pa.: Dowden, Hutchinson and Ross, Inc., 1975.

Linton, D. The assessment of scenery as a natural resource. *Scottish Geographical Magazine*, 1968, *84*, 219–238.

Litton, R., Jr. *Forest landscape description and inventories—A basis for land planning and design. U.S.D.A. Forest Service Research Paper PSW-49, Pacific Southwest Forest and Range Experiment Station, Berkeley, Calif.*, 1968.

Litton, R. B., Jr. Aesthetic dimensions of the landscape. In J. Krutilla (Eds), *Natural environments*, Baltimore, Md.: Johns Hopkins University Press, 1972, 262–291.

Litton, R. B., Jr. *Landscape control points: A procedure for predicting and monitoring visual impacts.* U.S.D.A. Forest Service Research Paper PSW-91, Pacific Southwest Forest and Range Experiment Station, Berkeley, Calif., 1973.

Litton, R. B., Jr. Visual vulnerability of forest landscapes. *Journal of Forestry*, July 1974a, *72*, 392–397.

Litton, R. B., Jr. Esthetic resources of the lodgepole pine forest. *Proceedings, Management of Lodgepole Pine Ecosystems Symposium October 9–11, 1973* Pullman, Washington, 1974b.

Lovejoy, D. *Land use and landscape planning.* New York: Barnes and Noble, 1973.

Lucas, R. C. *The recreational capacity of the Quetico-Superior Area.* U.S. Forest Service Research Paper LS-8, 1964.

McHarg, I. *Design with nature.* New York: Natural History Press, 1969.

Mills, L. V., Jr. *Lorne scenic quality study.* Urbangroup, Lorne, Australia, January, 1975.

New England Natural Resources Center, *New England Natural Areas Project*, Boston, Mass. 1974.

Palmer, J., and Zube, E. H. Numerical and perceptual landscape classification. In E. H. Zube (Ed.), *Studies in landscape perception*, Institute for Man and Environment, University of Massachusetts, Amherst, 1976.

Rapoport, A. An approach to the construction of man environment theory. *Enviromental design research.* Vol. 2. Stroudsburg, Pa.: Dowden, Hutchinson and Ross, Inc. 1973, 124–135.

Riotte, J., Fabos, J. G., and Zube, E. H. Model for evaluation of the visual–cultural resources of the Southeastern New England Region. In E. H. Zube, R. O. Brush, and J. G. Fabos (Eds.), *Landscape assessment: Values, perceptions and resources.* Stroudsburg, Pa.: Dowden, Hutchinson and Ross, Inc. 1975.

Saarinen, T. F. *Perception of Environment.* Washington, D.C.: Association of American Geographers, 1969.

Saarinen, T. F. Environmental perception. In I. R. Manners and M. W. Mikesell (Eds.), *Perspectives on environment.* Washington, D.C.: Association of American Geographers, 1974.

Shafer, E. L., Jr. Forest aesthetics—a focal point in multiple use management and research. *Proceedings 141VFRO Congress*, Munich, 1967, Paper 7, Section 26.

Shafer, E. L., and Thompson, R. C. Models that describe use of Adirondack Campgrounds. *Forest Service*, December 1968, *14*, 383–391.

Shafer, E. L., Hamilton, J. E., and Schmidt, E. A. Natural landscape preferences: a predictive model. *Journal of Leisure Research*, 1969, *1*, 1–20.

Shafer, E. L., and Mietz, J. *It seems possible to quantify scenic beauty in photographs.* U.S.D.A. Forest Service Research Paper NE-162, Northeast Forest Experiment Station, Upper Darby, Pa., 1970.

Shafer, E. L., and Tooby, M. Landscape preferences: An international replication. *Journal of Leisure Research*, 1973, *5*, 60–65.

Smithsonian Institution. *Planning considerations for statewide inventories of critical envi-*

ronmental areas: A planning guide (Report 3). Washington, D.C., Center for Natural Areas, Office of International Environmental Programs, September, 1974.

Stanford Research Institute. *Aesthetics in environmental planning.* Washington, D.C.: Environmental Protection Agency, EPA-600/5-73-009, U.S. Government Printing Office, November, 1973.

Stanky, G. A strategy for the definition and management of wilderness quality. In Krutilla, J. (Ed.), *Natural environments.* Baltimore, Md.: Johns Hopkins University Press, 1972, 88–114.

Twiss, R. H. Planning for areas os significant environmental and amenity values. In D. M. McAllister (Ed.), *Environment: A new focus for land-use planning.* Washington, D.C.: RANN, National Science Foundation, U.S. Government Printing Office, 1973.

U.S.D.A. Forest Service. Title 2400-timber management. *Forest Service Manual, Portland, Oregon,* April, 1968.

U.S.D.A. Forest Service. *Management practices on the Bitterroot National Forest,* Missoula, Montana, 1970.

U.S.D.A. Forest Service. *National forest landscape management.* Vol. 1. Agriculture Handbook 434, Washington, D. C.: U.S. Government Printing Office, February, 1973.

U.S.D.A. Forest Service. *National forest landscape management.* Vol. 2. Agriculture Handbook 462, Washington, D.C.: U.S. Government Printing Office, April, 1974.

U.S. Senate. *A university view of the forest service.* Committee on Interior and Insular Affairs, Senate Document No. 91-115, 1 December 1970.

Veal, A. J. Perceptual capacity: A discussion and some research proposals. Working Paper No. 1. Center for Urban and Regional Studies, University of Birmingham, England, March, 1973.

Water Resources Council. Water and related land resources, establishment of principles and standards for planning. *Federal Register,* Washington, D. C., *38,* 174, Part III, Monday, 10 September 1973.

Wright, G. McK. Landscape quality: A method of appraisal. *Royal Australian Planning Institute Journal,* October 1973, 122–130.

Zube, E. G. *The islands, selected resources of the United States Virgin Islands and their relation to recreation tourism and open space.* University of Massachusetts, Amherst (for the U.S. Department of the Interior), 1968.

Zube, E. H. Evaluating the visual and cultural environment. *Journal of Soil and Water Conservation,* July–Aug. 1970, *25:4,* 137–141.

Zube, E. H. Scenery as a natural resource: Implications of public policy and problems of definition, description and evaluation. *Landscape Architecture Quarterly,* January 1973a, 126–132.

Zube, E. H. Rating everyday rural landscapes of the Northeastern U.S. *Landscape Architecture Quarterly,* 1973b, *63*(3), 370–375.

Zube, E. H. Cross-disciplinary and intermode agreement on the description and evaluation of landscape resources. *Environment and Behavior,* 1974a, *6*(1), 69–89.

Zube, E. H. An alternative strategy for land use planning. In *53rd Annual Meeting Proceedings, Appalachian Section, Society of American Foresters,* Greensboro, North Carolina, 1974b, 14–25.

Zube, E. G., and Carlozzi, C. *Selected resources of the island of nantucket.* Publication No. 4, Cooperative Extension Service, University of Massachusetts, Amherst, 1966.

Zube, E. H. *Study of the visual and cultural environment,* North Atlantic regional water resources study. Research Planning and Design Associates, Inc., Amherst, Mass. (for the NAR Study Coordinating Committee) 1970.

Zube, E. H., and Fabos, J. G. *Environmental base study*. Southeastern New England Study of Water and Related Land Resources, New England River Basins Commission, Boston, 1972.

Zube, E. H., and Isgur, B. Agency-university-local cooperation in natural resources planning. *Journal of Soil and Water Conservation*, September–October, 1975.

Zube, E. H., and Mills, L. V. Cross-cultural explorations in landscape perception. In E. H. Zube (Ed.), *Studies in landscape perception*, Institute for Man and Environment, University of Massachusetts, Amherst, 1976.

Zube, E. H., Brush, R. O., and Fabos, J. G. *Landscape assessment: Values, perceptions and resources*. Stroudsburg, Pa.: Dowden, Hutchinson and Ross, Inc., 1975.

Zube, E. H., Pitt, D. G., and Anderson, T. W. *Perception and measurement of scenic resources in the southern Connecticut River valley*. Institute for Man and Environment, University of Massachusetts, Amherst, 1974.

Zube, E. H., Pitt, D. G., and Anderson, T. W. Perception and prediction of scenic resource values of the Northeast. In E. H. Zube, R. O. Brush, and J. G. Fabos (Eds.), *Landscape assessment: Values, perceptions and resources*. Stroudsburg, Pa.: Dowden, Hutchinson and Ross, Inc., 1975, 151–167

4

Motivational and Social Aspects of Recreational Behavior

DAVID C. MERCER

Over the past 15 years or so a number of countries around the world—notably the United States, Canada, Sweden, and Great Britain—have expended a great deal of time, finance, and technical and bureaucratic energy in the execution of sophisticated national recreation surveys, designed to identify participation rates in a variety of (primarily) outdoor activities, the trends in participation, and the factors influencing leisure time behavior. The main rationale for these large-scale empirical surveys invariably has been twofold: to chart in a quantitative fashion the dimensions of the recreation "problem" and, where possible, model and predict future patterns of participation. By their very nature such investigations are concerned almost exclusively with large statistical aggregates and with average or "modal" behavior rather than with individuals or small groups. The emphasis consequently is on relatively easily measurable independent variables such as age, income, and occupation and dependent variables such as recreation days, visitor days, and the like. Likewise, the preoccupation generally is with analytical elegance or simplicity rather than with comprehensiveness, and with *prediction* at a gross social aggregate level rather than with *understanding* the leisure phenomenon at the scale of the individual or small group. Findings are generally reported

DAVID C. MERCER · Monash University, Melbourne, Australia.

on a national or broad regional basis, and major financial commit-
ments are often made on the basis of the survey results. For example,
the finding of the Outdoor Recreation Resources Review Commission
(ORRRC) that driving for pleasure is the most frequently engaged-in
activity in the United States resulted in a proposal to spend 4 billion
dollars on a national system of scenic roads in that country.

Although aggregate recreation systems modeling is still being
carried on at a frenzied pace at national and regional scales in many
parts of the world, a number of years have elapsed since the first
large-scale surveys were instituted and the results analyzed, and
recently several writers have begun to look very critically at the
methodologies, parameters, and results of these investigations.
Brown, Dyer, and Whaley (1973, p. 16) argue forcibly that "most
recreation research to date cannot stand the question, 'So what?'," and
that the field "has largely ignored the broader social context of the role
of recreation in satisfying man's needs or solving problems of the
appropriate role of recreation in competing among alternative uses of
resources" (p. 17). LaPage (1971, 1972) has scathingly attacked the
usefulness of the "mythical" generalizations of the various national
surveys, and Hendee (1971), too, has put a strong case for more
relevant, policy-oriented research in the leisure field.

Basically, the problem lies in the extreme complexity of the leisure
phenomenon (Appleton, 1974; Murphy, 1974). One frustrated social
scientist claimed that attempting to conceptualize and analyze leisure
behavior was like attempting to grab hold of a jellyfish with one's bare
hands; and Ennis (1958, p. 259) has written that "the multiplicity of
possible leisure goals and especially their lack of familiar ordering
gives what amounts to an unpriced cafeteria selection quality to
leisure and thus enhances its shifting and problematical character."

Leisure behavior does not fit easily into the social scientist's
precoded questionnaire format. Both Wolfe (1970) and Cichetti, Seneca, and Davidson (1969) have demonstrated how formidably sophisticated and unwieldy a recreation systems model needs to be before
one comes close to a reasonable level of prediction, since so many of
the parameters—time/distance and attractiveness, for example—are
considerably modulated by the perception of the recreationists concerned. Similarly, Chubb (1971) has drawn attention to the fact that
many of the questions framed by white, middle-class researchers are
frequently totally outside the comprehension of the respondent. Questions are either asked in the wrong way, they force the interviewee to
be precise when he sees only vagueness, complexity, and irregularity,

or the researcher inquires after only a very narrow range of recreational activities. In real life many leisure pursuits are taken up for a while and then promptly dropped, and outdoor sites are frequently visited on a highly irregular basis. Moreover there appears to be a very large nonparticipating public that remains virtually ignored by researchers, and there is very little correlation between the "traditional" socioeconomic parameters so endearing to social scientists and observed recreational behavior. Depending upon the study that is being quoted, something between 5% and 30% only of the variation in participation rates is "explained" by reference to socioeconomic attributes. The Outdoor Recreation Resources Review Commission Study, for example, concluded that "it follows that characteristics additional to the [nine] socioeconomic characteristics included in the analysis are *major* determinants of levels of outdoor recreation activity" (Study Report 20, 1962, p. 69; author's italics). The report then went on to suggest a number of additional variables such as experience in childhood, preferences of other family members, and individual leisure goals and values, which it deemed would be worthwhile investigating in this context. This approach would seem to be a fruitful one for, as the nationwide median American income fast approaches $10,000, income—previously a major determinant of differences in recreational behavior—is rapidly becoming redundant as a behavioral indicator (Burdge and Field, 1972). On this point Havighurst (1958, p. 182) has written that "if the difference in income between manual and nonmanual workers becomes less, then the main differences between classes will be non-economic, and the nature of these differences will be largely determined by the uses they make of their money and of their leisure time." Often the relevant questions for the investigator now are not so much *whether* a person owns an automobile or a boat, but *what kind* of automobile or boat? Then, once that question has been answered, the research emphasis shifts naturally to a consideration of the individual preferences and constraints influencing the use or nonuse of that particular leisure "good."

The purpose of the present chapter, then, as in previous publications (Mercer, 1971, 1973, 1975), is to shift attention away somewhat from the broad, highly aggregated social level of recreation research (represented by the sophisticated "predictive" systems modeling approach) by emphasizing more strongly the social psychological level of individual or small-group leisure-time behavior. The main aim of the discussion is to point the way toward a greater understanding of leisure behavior in general and the individual recreational experience in particular. That such an approach is essential is well recognized by Clawson (1972, pp. 59–60):

> Increasingly, what we have called resource management in the past may become people management in the future. . . If people are to be "managed" or at least influenced in their direct use of natural resources, then resource managers will have to know much more about people, their motivations, their sensitivities, and their responses to various stimuli.

The discussion is ordered in terms of scale. It starts with a consideration of the rather general societal forces operating on individuals and influencing both decisions to engage in a particular recreational pursuit or range of pursuits and environmental preferences. Place of residence is given special attention here. Then the focus is on the more direct social forces playing on the individual; and, finally, attention is shifted to the psychological level. The terms "leisure" and "recreation" are used interchangeably throughout the chapter to refer to "nonwork behavior in which people engage during free time" (Murphy, 1974, p. 109). The recreational activities that will be considered are essentially those informal, nonsporting pursuits such as pleasure-driving, boating, sightseeing, hiking, swimming, and camping that take place outdoors and generally, though not exclusively, on public land. The chapter is multidisciplinary in scope with studies discussed from the fields of sociology, psychology, anthropology, economics, social psychology, forestry, and geography.

MOTIVATIONAL ASPECTS OF RECREATION

Why do individuals engage in one recreational activity rather than another? How do their activity preferences change through the life-cycle? Do persons engaging in one activity also engage in a similar range of other recreational pursuits? Why are some individuals extremely active in their leisure time while others are quite clearly nonparticipants? How do we account for varied environmental preferences? These are some of the broad issues that will be taken up in this chapter.

The first point to be stressed is that leisure behavior in general is no different from other aspects of human behavior in terms of the motivational forces involved. The same psychological and sociological concepts and constructs are as relevant to the study of leisure as they are to investigations of such subjects as work or interpersonal relations. Indeed, in many instances it is extremely difficult, if not impossible, to clearly demarcate leisure or recreation as a separate field of activity and inquiry from other aspects of human behavior. What is work for one person is leisure for another, and an activity that is work for one person at one time may be leisure at another time. Reading, gardening, driving, and "do-it-yourself" activities around the home are all representative examples.

Table 1 demonstrates that for any given individual, depending upon the circumstances, the same activity can be considered either an essential or leisure act, thereby underscoring Clawson's point that "the distinguishing characteristic of recreation is not the activity itself but the attitude with which it is undertaken" (in Clawson and Knetsch, 1966, p. 6). In any objective sense most people would probably define shopping, for example, as a necessary act, but Hemmens (1970) has suggested that many multigood shopping expeditions, in particular, would be more satisfactorily defined as recreational excursions.

TABLE 1
A TYPOLOGY OF TIME
(After Cosgrove and Jackson, 1972)

	Fully committed (essential)	Partly committed (optional)	
		Highly committed	Leisure
Sleeping	Essential sleep		Relaxing
Personal care and exercise	Health and hygiene		Sport and active play
Eating	Eating		Dining and drinking out
Shopping	Essential shopping	Optional shopping	
Work	Primary work	Overtime and secondary work	
Housework	Essential housework and cooking	House repairs and car maintenance	Do-it-yourself, gardening
Education	Schooling	Further education and homework	
Culture and communication (nontravel)			TV, radio, reading, theater, hobbies, and passive play
Social activities		Child raising, religion, and politics	Talking, parties, etc.
Travel	Travel to work/school		Walking, driving for pleasure

In short, in order to attempt to understand an individual's leisure preferences and life-style it is first necessary to come to an understanding of that person's values and attitudes, which, in turn, are a function of a combination of personality and the encompassing social and physical setting (see Figure 1). All of these are in a constant process of change. The objective social and physical settings, the roles

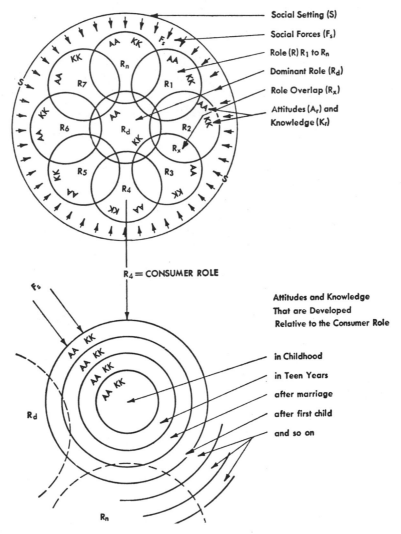

Figure 1. Human behavior and consumer behavior. (After McNeal, 1965.)

an individual plays, and the attitudes and knowledge of a person all change throughout the life cycle. But, more importantly, the perception of various social forces, the relative importance of different roles, and the perception of the physical setting can also—and frequently do—change markedly over time.

THE HIERARCHY OF NEEDS

Every individual has certain basic needs that he or she constantly strives to satisfy. A hierarchy of such needs was first outlined in detail in 1954 by Maslow in his book *Motivation and Personality*, and was later related more specifically to leisure behavior by Farina (1969) in an article entitled "Toward a Philosophy of Leisure." Basically, Maslow argued that needs are arranged in a hierarchical fashion ranging from low level to higher level needs, and that the lower level, or more basic needs have to be satisfied before the individual has the will or freedom to aspire to higher level wants. Starting with the low level needs first, Maslow's hierarchy of needs reads as follows:

1. *Physiological needs:* These include hunger, thirst, sex, activity, rest, homoeostasis, and bodily integrity. These are needs which once relatively well satisfied free the individual to address himself to the next set of needs.
2. *The safety needs:* In this group are the need for orderliness, justice, consistency, routine, predictability, limits, and physical safety. Adult expression of these needs is evidenced by interest in job security, savings accounts, and various insurance plans covering a variety of life's exigencies.
3. *The belongingness and love needs:* These needs emerge after both the physiological and safety needs are fairly well satisfied. They include the need to love and be loved, the need for friendship, interpersonal relationships, and a sense of identity with a group.
4. *The esteem needs:* There are two sets of esteem needs, self-esteem and the esteem of others. The former refers to the need for strength, achievement, adequacy, mastery, competence, and independence; the latter refers to prestige, reputation, status, dominance, recognition, attention, and appreciation.
5. *The need for self-actualization:* The emergence of this need usually depends upon the prior satisfaction of the needs at the four lower levels. It is the need for self-fulfillment, the need "to become everything that one is capable of becoming." (Farina, in Murphy, 1974, pp. 152–153)

Clearly, Maslow's classification represents an oversimplification of reality, but there is some evidence that it relates fairly closely to observed patterns of behavior. Both Turnbull (1974) and Bettelheim

(1970), for example, with reference to two quite different situations—a primitive East African tribe and the occupants of German concentration camps during the Second World War—have documented with terrifying clarity how the basic human needs for love, justice, activity, and the like are affected at both the individual and societal level under conditions of extreme stress.

Farina argues that once the individual has been freed from the necessity of striving to satisfy the lower level needs, he is liberated to quite literally play at the more basic wants. Sampling a variety of foreign foods in restaurants, dressing up and reenacting historical battles, joining social clubs or challenging orderliness, safety, or predictability by gambling or mountaineering are all examples of play activities that have their more serious counterparts lower down the needs ladder. Change the time, change the social situation, change the attitude, and the freely chosen risk-taking behavior of the mountaineer, canoeist, or undersea diver becomes a matter of necessity.

Apart from the basic necessities of food and shelter, etc., needs are culturally and socially defined. Burch (1974) notes that "each culture is a distinctive language which shapes the perceptions and behavior of its members" (p. 61), and he stresses the very marked differences in the perception of such basic concepts as time and work between preindustrial societies such as the Tikopia and contemporary America. Burch argues that the fact that leisure is viewed as a problem in affluent societies can be traced directly to a number of cumulative human inventions such as money, urbanism, industrialism, the work ethic, and an abstract conception of time that radically influence the way people in industrial nations view their various activities:

> It makes a good deal of difference in one's orientation to life if one has a conception of minute divisions ever dribbling away, or whether one sees life divided into the broad time span of seasons with their attendant problems and ritualized celebrations. (Burch, 1974, p. 68)

Burns (1973) developed this theme in the context of industrialization in nineteenth-century Great Britain. He pointed out that the commercialization of leisure proceeded hand in hand with industrialization. As work increasingly came to be something done in a special place at a certain, specified time, so leisure increasingly came to be the vacuum between work slots, a void that, initially, was largely filled by the commercialized entertainment business. As is apparent, this phenomenon has since grown to mammoth proportions in the form of commercialized sport, the film, television, pop music world, and the package tour vacation business, to name but a few examples. Today there exists no aspect of recreation that is not commercialized to some degree or other. In short, "Social life outside the work situation. . .has been created afresh, in forms which are themselves the creatures of industrialism, which derive from it *and which contribute to its development, growth and further articulation*" (Burns, 1973, p. 46; author's italics).

It follows that advertising, the conscious creation and manipulation of needs, has come to be strongly associated with recreation and leisure goods and activities in the same way that it is associated with the sale of any commercial product. There now exist literally thousands of magazines devoted to the dissemination of information relating to hundreds of individual recreational pursuits ranging from power-boating to squash. Similarly, the extent to which the advertising agency industry has come to favor natural outdoor environments and a variety of recreational settings as a backdrop for its promotional films and photographs is now well appreciated (Coughlin and Goldstein, 1968). The influence of the mass media in the leisure educative process is obviously enormous. Newspapers, magazines, films, and radio and television constantly bombard the public with information about new trends in recreational behavior, with ideas for places to visit and with articles on recreational equipment (Kaplan and Lazarsfeld, 1962).

It would be surprising if the sheer volume of such material did not have a very pronounced influence on recreational behavior, but so far the extent of this impact has eluded precise analysis. Intuitively, though, it would appear that the media are influential in at least two main ways. In the first place, access to information about an activity or leisure good must have an effect, either positively or negatively, on the aspirations of people with regard to that pursuit or item of leisure equipment. The widespread ownership of television by blacks in the United States, for example, would appear to be a significant factor in changing black recreational patterns (Craig, 1972). The same generalization obviously applies to white people also. Kaplan and Lazarsfeld (1962, p. 207) stress that television is an instrument of democracy "providing the possibility, through symbols of movement, of a psychological life away from the community." Projecting into the future, they also propose that "its second phase may well be to serve as the transition to a real as well as to nominal experience." Second, it would seem that the need for the media to be constantly novel and newsworthy, continually presenting the public with new images and ideas, could well have its expression in the consumer behavior area in the faddish nature of many recreational activities, which—like the latest "in" book or film—are taken up for a while and then promptly dropped (Linder, 1970). Indeed, when one considers the powerful and pervasive influence of the media and the opinions of friends and relatives on the "image" of various recreational pursuits, the question could well be put: Is anyone ever totally free to choose his or her own leisure-time activities? In view of the fact that freedom of choice is generally taken to be the distinguishing characteristic of leisure, this would seem to be a very pertinent question (see Kerr, 1965).

Table 2 illustrates the transformation of leisure that takes place as one moves from preindustrial folk and feudal societies toward the industrial, and eventually postindustrial societies. Each ideal-type society presents the individuals living within it with certain opportunities and possibilities for recreational participation, but the activities and associated environments that are perceived as possible or open are, again, markedly influenced by the specific subculture within which a person lives. In order to understand shifts in recreational participation that are taking place at the individual level we must constantly bear in mind the immensely powerful social, economic, and technological forces that are operating at the broad societal level (Kaplan and Bosserman, 1971). To take a highly simplistic example, the recent energy crisis had a noticeable impact on participation in bicycling in Western Europe and North America. A prolonged and major domestic energy crisis would obviously have a dramatic effect on participation in a large number of recreational activities around the world. On this question Brown (1970) has stressed that at the individual level economic considerations are frequently of great significance in deciding to what extent people will engage in particular recreational activities. He makes use of the concept of consumer sentiment, which refers to the consumer's perception of general market conditions, the economic outlook for the future, and the individual's personal financial situation. Brown suggests that the timing of national surveys of recreational participation will markedly influence the results produced since consumer behavior closely reflects the ups and downs of the economy. He presents tentative evidence from the 1960 and 1965 Bureau of Outdoor Recreation National Surveys to verify his point, and concludes that ". . .if recreation acts like other discretionary purchase products, part of the increased participation rates in 1965 were likely due to recreationist (consumer) perceptions of good times" (Brown, 1970, p. 265).

The emphasis so far in this discussion of recreational motivation has been on those pervasive, often hidden, broad external societal forces impinging on an individual and markedly influencing the

TABLE 2

TRANSFORMATION OF CULTURE TOWARD A SOCIETY OF LEISURE
(AFTER MURPHY, 1974)

	Folk society	Feudal society	Industrial society	Postindustrial society
Community life	Small, isolated nonliterate, homogeneous communities with a strong sense of group solidarity. No distinction of class. Social structure is rigid. Mobility is slow and infrequent. Organic, human relationships.	Relatively large peasant population and small elite. More stratified and more heterogeneous. Small core of literate persons, priestly class. Some government apparatus. Integration of all phases of daily life for aristocracy.	Large community size and relatively dense population. Heterogeneity of people and cultures. Anonymity, transitory and impersonal relationships. Social mobility. Fluid class system, mass literacy. Predominance of secondary contracts.	Shared community decision making based on pluralistic cooperative relationships.
Work life	Energies wholly oriented toward the quest for food. Little or no specialization of labor. No food surplus. People produce their own artifacts.	More occupationally differentiated than folk communities. Trade commerce, and craft specializations well developed.	Occupational specialization, division of labor. Economy of scarcity dominated by manual labor, assembly-line work.	Economy of abundance, economic independence. Still large work force dominated by science and technology. Growing number of craftsmen and artisans characterized by individualistic, nonmachine, and stylistic qualities.
Free time	Behavior is traditional, spontaneous, spiritual, personal. Sacred prevails over secular. Leisure is part of living, condition interwoven into the main fabric of life.	Leisure available to an elite upper class, integrated into the rituals, celebrations, weddings, and day-to-day routines of the masses.	Mass leisure. Leisure a specific block of time, earned from work. Socially acceptable leisure behavior predominates.	Individualized and liberated leisure based on inherent right and specified by particular individual needs. Many personal options, diverse styles of life. Fusion of work–leisure relationships.

configuration of his or her needs. As I have stated elsewhere (Mercer, 1973), needs can be felt, but these feelings need not necessarily be translated into action, or behavior. In order for the latter to occur, certain preconditions are necessary: The individual has to be able to articulate to himself what his needs are, and an acceptable recreational opportunity that the recreationist is aware of has to be available and accessible. Thus, the scale of the discussion will now be shifted somewhat toward a consideration of the rather more personal forces imposing on an individual and influencing his or her motivations. We will begin with a consideration of the importance of the *place of residence* variable. Then the emphasis will be on the influence of the very powerful and direct *social forces* of relatives, friends, and work mates. And then, finally, attention will be focused on *personality* as a factor in leisure motivation.

THE IMPORTANCE OF PLACE OF RESIDENCE

Place of residence or, more accurately, an individual's residential history is an immensely important factor influencing recreational preferences and needs. Farina (1965, p. 4), for example, has written that "...for each person the norms, folkways and values of his specific locale strongly influence and limit his range of leisure choice." Any given place of residence, of course, may be viewed within an almost infinite number of spatial frameworks of different orders of scale. In a very general sense the national setting is of relevance to the current discussion, for topography and climate, the size, density, and distribution of the population, the wealth and historical development of the country, the administrative structure and cultural attitudes to resources—all these, together, contribute toward a given pattern of supply opportunities and, some would argue, a unique national personality (see Hollander, 1969). At the more localized scale, regions vary in their recreational resource endowment and this, perhaps more than anything else, has a direct impact on the manifest recreational behavior patterns of the local population. In commenting on regional variations in participation rates in different activities in the United States, Knetsch (1967, p. 7) noted that

> people make greater use of water recreation facilities per capita in the Maritimes than they do in the Prairies, the differences having more to do with the availability of water than with differences in income, education or age distribution between the two populations.

Similarly, as Baumann (1969) so clearly demonstrated with reference to the recreational use of domestic water-supply reservoirs in different parts of the United States, there is evidence that consumers simply adapt unquestioningly to variable supply situations. In those areas of the country where recreational use was forbidden on water-supply reservoirs recreationists did not use them, apparently accepted the logic of the ruling, and presumably substituted with alternative recreational sites and/or activities. The importance of such substitution behavior will be taken up again later in the chapter. However, for the purposes of the present discussion it is the real or supposed differences in recreational behavior between rural and urban dwellers that are of particular interest. There now exists a sizable body of literature on the question of whether or not urban dwellers display a characteristic pattern of outdoor recreational behavior that marks them off as different from rural inhabitants.

Hendee (1969a) summarized the results of the 1960 National Survey relating to the place of residence variable and found that for only two out of 18 activities, fishing and hunting, were rural dwellers overrepresented among participants (Table 3). These results would seem to indicate that the outdoor recreation environment—in particular, the relatively unspoiled rural outdoor recreation environment—is especially attractive to the urban dweller, perhaps because it allows him an experience that cannot easily be found in an urbanized setting.

Three motivational theories of leisure behavior have been advanced in relation to this and other questions, namely, the *compensatory*, *familiarity*, and *personal community* (or *pleasant childhood memory*) theories (Burch and Wenger, 1967; Burch, 1969). Let us look at each of these in turn.

The basis of the compensatory hypothesis is that

> whenever the individual is given the opportunity to avoid his regular routine he will seek a directly opposite activity. The idea implies a safety-valve effect; whenever boredom or monotony of routine builds up the individual requires a sharply different experience or he will break down (Burch, 1969, p. 127).

It is clear that this hypothesis is an implicit assumption underpinning much writing seeking to explain outdoor recreational behavior. Green (1964), for example, has argued that in their leisure time many contemporary Americans seek to recapture a sense of communion with nature that formerly characterized early settlement on the continent. Hendee (1969a, b), too, has suggested that urban living, in locations divorced from the natural environment, results in the development of an appreciative rather than utilitarian attitude toward nature. An appreciative value position is represented, for example, by participa-

TABLE 3
PARTICIPATION IN SPECIFIC OUTDOOR RECREATION ACTIVITIES BY RURAL
VERSUS URBAN RESIDENCE
(AFTER HENDEE, 1969a)

Activity	Rural over represented	Urban over represented	Some variation in trend	Percentage of population participating in 1960
Outdoor games and sports		X		30
Bicycling		X		9
Horseback riding		X		6
Ice-skating	X		X	7
Sledding and tobaganning		X		9
Snow skiing[a]		X		2
Swimming		X		45
Boating		X	X	22
Water-skiing		X	X	6
Fishing	X			29
Camping	X		X	8
Hiking		X	X	6
Hunting	X			11
Picnicking[b]	X	X		53
Walking for pleasure		X		—
Nature walks	X	X	X	14
Driving for pleasure[c]		X		52
Sightseeing		X	X	42

Source: Condensed from text in ORRRC *National Recreation Survey*, Study Report No. 19, 1962, pp. 8–54.
[a] No analysis of residence but presumed to be urban.
[b] Curvilinear relationship from urban—small urban—rural nonfarm—rural farm with peaks in use by urban and rural nonfarm.
[c] Curvilinear relationship with participation highest in small urban places and lower in large urban and rural areas.

tion in remote wilderness-type activities, while compensation for rural dwellers may be found in busy, exciting urban settings. Thus, the ORRRC Study Report on wilderness use concluded that ". . .city people are more likely to escape to the country for their vacations while those born (or living) in the country are more likely to be pulled to the attractions of the city" (ORRRC Study Report 3, 1962, p. 136). The compensatory theory suggests, then, a society constantly seeking new leisure-time experiences in novel environmental settings. In particular, it points to people having a leisure style that is diametrically opposed to their occupational life-style. Taking the example of camping, Burch (1969, p. 134) found that "those with 'soft' patterns of physical labor seem more likely to seek the isolation and exertion of remote camping; while those with 'hard' patterns of physical labor seek the ease and sociability of easy access camping."

In contrast, the familiarity theory, as its name implies, suggests that in their leisure time people engage in activities that are familiar to them and are not markedly dissimilar from their everyday lives. Hendee (1969a) has suggested that the harvesting recreational pursuits of hunting and fishing are overrepresented among rural dwellers because these pursuits closely relate to the utilitarian attitude to nature so prevalent among rural dwellers in their everyday lives; and Etzkorn (1964, p. 83), in concluding his camping study, noted that he "failed to detect leisure type activities which differ markedly from those that are generally characteristic of the home environments of the campers." In a seminal paper published in the ORRRC Study Report 22, Hauser (1962, p. 46) argued that as urbanism becomes more widespread the demand for outdoor recreation will diminish. In other words, the "back to nature" movement will subside, and there will be a decreased demand from urban dwellers for access to unfamiliar natural environments. Hauser provided evidence from the 1960 National Survey to support his argument and Hendricks (1971) presented a similar case on the basis of his more recent surveys in Oakland and San Francisco.

The third, the personal community theory, assumes that an individual's leisure style is significantly affected by the social influence of close acquaintances, notably work mates, relatives, and friends. This hypothesis suggests that once a leisure style has been learned in childhood it tends to be carried through into adulthood. In other words, the father who as a child was introduced to a wilderness camping experience at an early age would be likely to continue with this form of activity and, in turn, introduce his own children to it. This particular hypothesis will be examined more closely in the next section of the chapter, but let us first take a critical look at the usefulness of the ideas that have been briefly outlined above.

The discussion has centered around the place of residence variable, with special attention being focused on the significance of the urban or rural residence of an individual as a variable influencing outdoor recreational preferences. The very least that can be said about this issue is that a review of empirical studies produces a very conflicting and confused picture that in no sense allows one to provide a clear-cut answer to this question. The case of wilderness recreation is illustrative of this confusion. The ORRRC study of wilderness use concluded that this type of recreation is most strongly favored by city born and bred urban dwellers for whom wilderness has a "compensatory" function to "relieve the anxieties and tensions of modern life" (Study Report 3, 1962, p. 151). By comparison, Burch's (1969) study of wilderness campers in Oregon found that the most primitive style of camping was linked with rural, not urban, residence, suggesting perhaps that wilderness users are simply following familiar routines.

We can be sure that place of residence is extremely important insofar as it presents an individual with a limited range of recreational opportunities (compare, for example, the recreational universe of the black slum-dweller with that of the affluent white middle-class surburbanite), but beyond that it is extremely difficult to isolate the effect of place of residence on activity preference structures. A detailed examination of a number of empirical studies conducted in this area leads one to suspect that frequently they are measuring different things and hence are not strictly comparable. Moreover, many of them appear to define both urban and rural in a very narrow, naive way, and to assume that activities labeled as hunting, fishing, camping, etc. are easily defined, homogeneous pursuits. The family, work, living, and leisure environments of an individual are complex, multidimensional entities, and all too rarely do researchers make it clear which specific dimension of which particular environment they are measuring. This being the case—aside perhaps from relatively hazardous activities such as rock-climbing—one could take virtually any recreational activity in any situation and argue that it supported one, or two, or all of the motivational theories reported above.

Burch (1969, p. 127) argues that compensatory behavior involves the individual seeking a directly opposite activity to his regular routine; but what is a directly opposite activity? Is white water canoeing the experiential opposite of high school teaching or taxi-driving? Is auto-camping the opposite of being a housewife? Obviously, in some senses these experiences are different from the average workday routine, but in other senses the experiences may be quite similar, or familiar. Everything depends on the individual's personality, work and social environment, and the activity in question. I would suggest that most recreational experiences, for most people, are partly compensatory and partly familiar. The housewife may well be carrying out the same cooking, cleaning, and child-minding activities in the auto-camping environment as at home, and the wilderness recreationist may be hiking a trail with the same set of familiar work colleagues. The settings are different but in an important social sense they are familiar.

It is not uncommon to meet or hear of individuals working in the cut and thrust of the business world who have "not had a vacation for years." Similarly, one comes into contact with people who quite easily find all their needs satisfied in the urban environment. Generally such people are not caught in boring or monotonous routines and so do not need the safety-valve effect of a recreational experience in natural settings. At the other extreme, many people in society *do* have a patently monotonous routine in their everyday lives and for various reasons are unable to take the opportunity to break out of it either temporarily or permanently, but they certainly show little evidence of "breaking down" as Burch suggests they should. Experimental evidence has clearly demonstrated that for the maintenance of normal, intelligent behavior people appear to require a continually varied sensory input (Bexton *et al.*, 1954), but for any given individual such variation can be acquired in a multitude of different ways. There is little evidence, for example, that the highly active person turned wheelchair invalid goes "mad." Rather, he tends to adapt to the inevitability of the situation and substitute with a totally different recreational behavior pattern (Gans, 1962; Hendee and Burdge, 1974).

Very little research in this area has attempted to come to terms with the complexity of the varied occupational, neighborhood, and leisure environments of individuals or with the multitude of substitute responses that are both theoretically and practically possible. Burch (1969) at least recognized that there are different styles of camping, (remote, auto, etc.), each with different meanings for the people concerned, and he also stressed that a person's residential history (as distinct from where they are currently living) is of importance; but he treated the occupational environment in a very gross manner. Burch's study was also concerned solely with those who participated in his three camping styles. This raises two questions: (1) To what extent are the different camping styles typical of the preferences of those responding to his particular interview schedule? (2) What generalizations are we to make about those people in the population with similar characteristics to Burch's interviewees who never go camping? Question (1) is of special significance in view of LaPage and Ragain's (1974) findings in relation to changing camping styles in the United States.

Knopp (1972) attempted to overcome some of these shortcomings in a more recent study in Minnesota. He interviewed people in their homes, thereby speaking with both participants and nonparticipants; and he attempted to operationalize the work and residence environments (but not the leisure environment) in a slightly more sophisticated manner than usual. For instance, a person's occupation was identified along such dimensions as crowding, supervision, variety, and general satisfaction. However, Knopp apparently failed to ask questions relating to residence history and he chose as his study area a county where his urban sample lived in a small town with a population of only 27,000. The question could well be asked: Is this really the ideal kind of situation in which to be looking for possible differences in recreational behavior between urban and rural populations?

To the writer's knowledge only one empirical study has ever attempted to assess substitution behavior in a quantitative fashion, and this is a recent investigation by Hecock and Rooney (1972). Their aim was to evaluate the leisure behavioral implications of the construction of a large new reservoir in Oklahoma. In other words, did the construction of the reservoir have any impact on, first, the activities that people in the catchment area engaged in? (Did the new reservoir, in effect, act as a motivating trigger to take up new recreational pursuits?) Secondly, did it influence the choice of locations that people favored for their recreation? Their results, based on before-and-after questions relating to comparative recreational behavior of those living within 100 miles of the reservoir in 1960 and 1970 are illuminating. Notwithstanding the considerable size of the reservoir (330 miles of shoreline), almost 40% of those interviewed indicated that neither they nor their families participated in any of the forms of outdoor recreation of interest to the researchers, a figure that is close to the national average in the United States. Then, of the recreationists, only about a quarter admitted that the reservoir had changed their recreational behavior patterns in any way. These findings underscore two important points:

1. The leisure environment of an individual encompasses a wide variety of readily substitutable activities and locations of which the outdoors is only one component (and perhaps a relatively unimportant one for large numbers of people).

2. It is possible that after a while an individual's recreational behavior becomes set in something of a habitual pattern with regard to favored activities and locations, and that a marked behavioral change only comes about as a consequence of increased crowding or for some other reason.

Maw and Cosgrove (1972) incorporated many of the above ideas in a hypothetical model of recreational behavior that clearly recognizes the complex substitutable nature of activities and locations, and also builds in the important factors of awareness and uncertainty (Figure 2). Maw and Cosgrove stress that at the scale of the individual

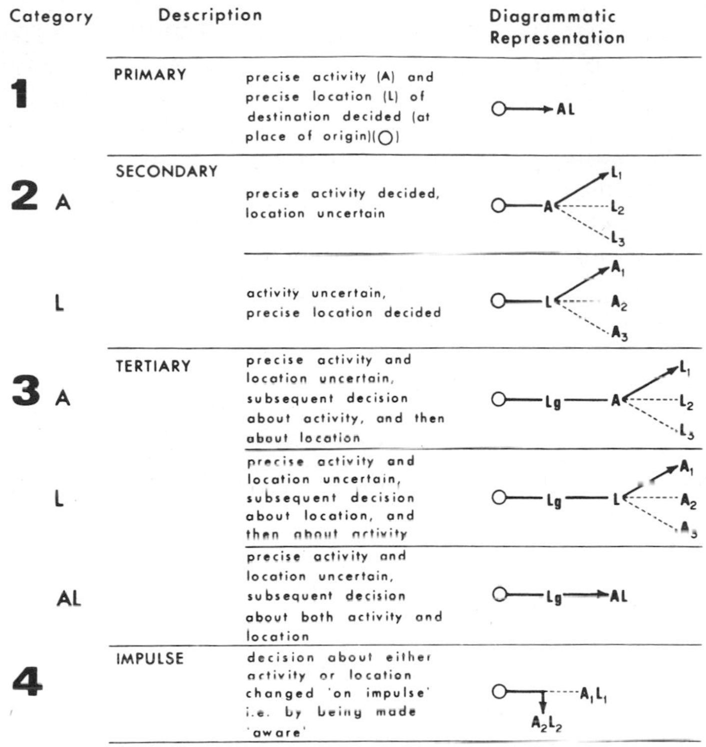

Figure 2. Awareness—activity and location linkages. (After Maw and Cosgrove, 1972.)

recreational trip (especially the day trip), it does not always follow that either the goal, the activity, or the location are clearly defined in the mind of the recreationist when he sets out. In some cases (category 1), to be sure, both the precise activity and location are decided in advance and the recreationist behaves according to a preplanned routine. But in many instances decisions about locations and activities are made en route. Consider the following hypothetical examples, which are intended to illustrate additional categories in Figure 2. In the case of category 2A in the diagram a decision may be made in advance to go surfing, but the precise location is undecided at the outset. A firm decision on this is only made later in the day after checking the wave conditions at a number of beaches. Similarly, in the case of category 4 (impulse behavior) a decision may be made to visit a certain state park for a picnic, but when the site is reached it is decided that it is too crowded and so a totally different activity and location is chosen for that day.

The important assumption in this model, as in that of Driver and Tocher (1974), is that recreational behavior should be viewed as a continually unfolding learning process rather than as an event or series of events isolated in time and space. Elson (1974) has suggested that when an individual or family moves to an entirely new residential location, their recreational trip-making behavior starts off as a more or less random search process and then only later settles down to a habit phase. His research indicated that four years may represent the typical length of the search phase in a new environment, though this time will obviously vary considerably in individual cases (see also Murphy and Rosenblood, 1974).

To conclude this section on the importance of the place of residence variable as a factor in recreational motivation, then, the following generalizations can, I think, be safely made:

1. The place of residence of an individual influences the kinds of recreational environments (and hence activities) that are potentially available to him.

2. There would seem to be some evidence suggesting differences between the recreational pursuits favored by those living in extremely rural environments on the one hand and extremely urban on the other; but that for people living in smaller urban places between these two extremes, much would depend on the size and/or location of the settlement in question. This suggests that it is meaningless to conduct interviews in an isolated urban place of 20,000 people set in a rural farming area and then generalize the results from this to all urban places as though these findings would be equally applicable to New York or Chicago.

3. Intuitively, place of residence per se would seem to be of rather less significance than the residential history of an individual. The person who has lived in a number of different places has the opportunity to engage in a variety of recreational pursuits, becomes aware of different recreational possibilities and, inevitably, becomes sensitized to differences between the environments he comes into contact with. To use Sonnenfeld's (1966) terminology he is the "non-native" who ". . .has lived in some other kind of environment" and hence "brings with him this environmental experience plus certain 'relict values' which condition his attitude toward space and land-scape" (pp. 77–78).

4. Finally, the unique social forces of friends, relatives, and acquaintances impinging on an individual have a powerful bearing on the evolution and development of his attitudes to work and leisure and on the form and direction of his recreational drives. In some cases

such attitudes can be linked with a spatial variable such as rural residence (Burdge, 1965), but in general this is becoming less universal for, as Mueller and Gurin (1962, p. 14) point out: ". . .many value and interest differences between city and country people are disappearing, and decreasing differences in outdoor leisure patterns would seem to be part of this trend." Let us now turn to a closer consideration of the effect of social forces on leisure motivation.

SOCIAL FORCES

The significance of intimate and immediate (rather than wider societal) forces has already been touched upon in the previous section when Burch's personal community theory was briefly examined. As a person moves through the life cycle he or she is influenced more or less strongly at different stages by numerous individuals and groups. Initially the life-style of close relatives—parents, brothers and sisters, etc.—is of paramount significance. Then friends and schoolteachers come into the picture, usually followed at a later stage by the influence of new peers, work colleagues, a spouse, and possibly children. As Figure 1 demonstrates, this move through the life-cycle is characterized by the adoption of a number of roles at different times. Dependent child, college student, wife, breadwinner, are representative examples.

Most recreation surveys include questions relating to the age of respondents and then go on to correlate age with stated preferences and participation rates. But to be wholly meaningful the age of an individual should be viewed in context. It should be related to the broader concept of life cycle (Bossard, 1960). The stage of life cycle is a complex function of the ages of the wife, husband, and children, together with the number of dependents.

Each life-cycle stage carries with it its own responsibilities, which act as constraints and/or drives in relation to manifest leisure behavior. At certain times, when the children are very young, for example, the demands of child rearing are so great and so time-consuming that the opportunities for leisure-time pursuits are severely limited. At this time the leisure activities of a family are likely to revolve around the preferences of the children, perhaps taking the form of easy access auto-camping or the occasional picnic in a state park. In this situation outdoor recreation often performs the important function of increasing family cohesiveness, especially if one or both of the parents are in the work force (West and Merriam, 1970). Moreover—and I consider this to be a very important point that has generally been overlooked in recreation research—in the family context (as well as in other group situations), the activity and environmental preferences of the individual adult members of the household are frequently subordinated to the needs of the children (Mercer, 1974). By way of illustration, the father whose stated preference and experience lies in wilderness recreation may well endure, and to a certain extent enjoy, a crowded beach experience for the sake of his children. In a word, his perception of that crowded beach is markedly influenced by social forces, by the role he is playing at the time of his visit; the beach is defined as attractive in a social sense (Lee, 1972). Elsewhere (Mercer, 1975), I have criticized much of the photo-choice research into environmental preference rating on the grounds that it fails to take into account such variations in the social context of hypothetical or actual recreational visits to the exhibited scenes. Thus, as Albrecht et al. (1972) argue, the measurement of attitude alone does not necessarily lead to any useful generalizations regarding action because "the strength of the attitude–action relationship is contingent upon factors which, in addition to the attitude, are present in the behavioral situation" (p. 149).

At any given stage of the life cycle it is quite possible, and not at all uncommon, for an individual to have a number of different roles. In turn, each role may have quite different social influence and leisure-time pursuits associated with it, each activity being motivated by different forces. Sometimes, too, the same leisure activity may satisfy a number of divergent needs. The act of buying a boat or second home, for example, can be motivated by a number of quite different though convergent forces, and the subsequent use of the boat or second home can likewise be motivated by quite different drives, depending on the circumstances. At times the motive may be primarily one of status seeking; at other times it may be escaping from the city, socializing with the family, or even working.

It seems probable that the real (as oppose to the professed) motives for engaging in many recreational pursuits or purchasing leisure goods are often highly complex and obscure and not easily uncovered by asking the respondent to check one or more motivational statements on a questionnaire form. For instance, in affluent Western achievement- and consumer-oriented societies leisure is frequently used as a referent for status, with the result that many people participate in activities that they cannot really afford. As Kaplan notes, "Golf, winter vacations and travel abroad have in many individual cases become a style of life before they could be comfortably paid for by their consumers" (1968, p. 94). This is not to suggest that status

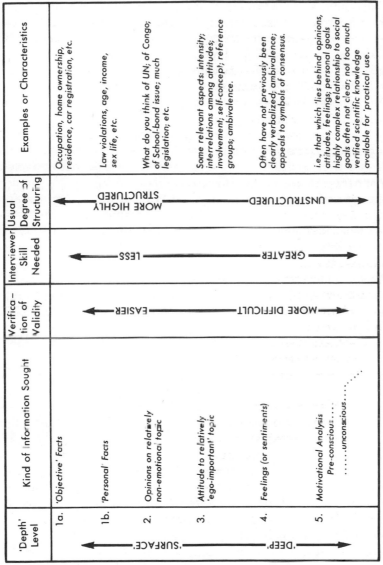

Figure 3. The relationship between type of information sought and interview structure, depth, verification, and skill. (After Crapo and Chubb, 1969.)

seeking is a motive in all such cases, but the point is that even if it is, the recreationist is unlikely to admit that this is so. Likewise, Leigh's (1971) detailed in-depth survey of the leisure behavior of 157 young unmarried adults in a northern English town suggested very strongly that their invariant leisure pattern of "progress from home to pub, to club, to an occasional dance and perhaps a football match, as a spectator, on Saturdays" (Rodgers, 1972, p. 224) was guided more strongly by the rules of the courtship ritual than by any other motivating force. The point is that in common with the investigation of other aspects of human behavior, recreation research can obtain facts that are either relatively superficial or, alternatively, deep (see Figure 3). So far, most leisure researchers have concentrated on the collection of relatively uncomplicated surface facts relating to such details as activity occasions in a given year or the distances traveled to national parks. However, the investigation of leisure motivation or the meaning of different recreational pursuits is far more complex and, I would argue, requires the application of techniques of the broadly defined participant-observation variety rather than (or in conjunction with) more highly structured methods (Campbell, 1970; Schatzman and Strauss, 1973).

We have already noted that for any given individual a constellation of roles may result in a number of varied recreational pursuits being carried on at the same stage of the life cycle. The businessman, for example, may have one or more activities that he engages in with his business clients, another group of activities he enjoys with his family, and yet another set that he participates in with his friends or on his own when the opportunity presents itself. Within recent years a considerable research effort has gone into the identification of typical groupings of recreational activities such that one can say with some degree of probability if person (x) engages in this type of recreational activity (y), it is also extremely likely that he will engage in this other recreational pursuit (z). The first serious effort in this direction was made by Proctor (1962). On the basis of a factor analysis of the ORRRC data he sought to group various recreational activities that apparently "go together." His four-factor grouping is illustrated in Figure 4, in which Proctor identifies passive, water, active, and backwoods factors linking various groups of activities.

More recently, both Tatham and Dornoff (1971) and Burton (1971) have used factor analysis and cluster analysis to group activities in the same way. Burton's data consisted of 1,056 respondents, each of whom was registered as having participated or not participated in each of 71 separate recreational activities. Cluster analysis was then carried out on these observations and 14 groups of activities were produced. The

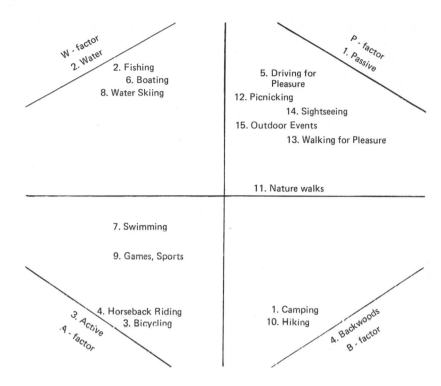

Figure 4. Factor structure of leisure activities. (After Proctor, 1962.)

following characteristics were suggested as being unifying variables in clustering activities together into certain relatively homogeneous groups: (1) skill; (2) group activity; (3) physically active; (4) risky or dangerous; (5) urban setting; (6) expensive for participants; (7) water-based; (8) speed, dexterity, rapid reactions.

In all such studies the assumption is made that the activities are typical of the respondents in the sense that they participate more or less regularly. This is a dangerous assumption, for participation in any given activity can range anywhere along a continuum from the once-only attempt through the ephemeral-role category to that of a central life interest bordering on fanaticism (Dubin, 1956; Devall, 1973; Steele and Zurcher, 1973). In addition, the individual activity classifications are highly aggregated and hence of little value for recreational planning purposes. We have already outlined above how camping is differentiated into a number of styles, each with varied meanings and reference groups for the individuals concerned (see also Clark, Hendee, and Campbell *et al.*, 1971); and the same argument is applicable to all recreational pursuits. Devall (1973), for example, has clearly demonstrated how the "conquest" activities of surfing and mountain-eering have a highly differentiated meaning structure for varied participant groups, and Lucas (1964) in his now classic research in the Boundary Waters Canoe Area demonstrated how segmented is the water factor of Proctor's typology.

Hendee *et al.* (1971) have been especially critical of such grouping exercises, arguing that they take account only of highly aggregated observed patterns of behavior and thereby ignoring latent preferences or the recreationist's degree of satisfaction with the activity or location in question. In short, such investigations group activities together in clusters that in reality may be quite meaningless for predictive and explanatory purposes. Accordingly, the authors have developed a typology of recreational activities that classifies user groups on the basis of stated preferences for activities and on the meaning structure of those activities rather than on observed behavior patterns. They suggest a fivefold classification of activities based on the attitudes of recreationists toward various pursuits as follows:

1. *Appreciative–symbolic*: Activities falling in this category are directed primarily toward the appreciation of the natural environment. The preservation of such an environment is thus essential for the maximum enjoyment of activities such as hiking, mountain-climbing, photography, and so forth.

2. *Extractive–symbolic*: This category refers to such activities as fishing and hunting that share many of the attributes of the first group but that are mainly oriented toward the capture of "trophies" from the natural environment. Burch (1965), in a somewhat similar classifica-tion, refers to such pursuits as "symbolic labor activities."

3. *Passive free play*: Included in this type are activities such as driving and sightseeing from a car, sunbathing, and quiet boating and canoeing. The main feature of such pursuits is that they require little physical effort; they are easygoing and unstructured activities.

4. *Sociable learning*: The distinguishing characteristic of this group of activities is that social interaction is the primary aim. Meeting people, talking with them, listening to lectures in a group—these are the kinds of activities subsumed under this heading.

5. *Active–expressive*: This type, which combines Burch's expres-sive and organized play categories, includes such active pursuits as swimming, water-skiing, and boat-racing.

Hendee and his associates suggest that, depending largely upon level of education (which they regard as a critical preference-forming

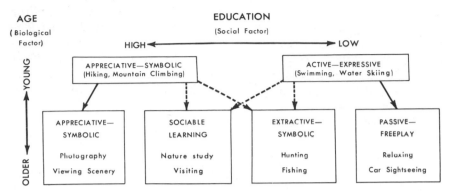

Figure 5. Hypothetical changes in activity preference with age and education. (After Hendee *et al.*, 1971.)

parameter), changes in activity preference could be expected to unfold through the life cycle approximately in the manner set out in Figure 5.

It is clear that these five groupings are far from discrete. To take an example: While wilderness hiking, for many participants, almost certainly has a dominant core dimension that may be labeled "appreciative symbolic" this in no sense negates the fact that the activity may also be characterized to differing degrees by aspects of "sociable learning." The authors are far from clear exactly how these dimensions are to be identified and analyzed.

However, the contrast between the work of Hendee and his colleagues and that of Proctor does point up the important fact that recreational activities can be classified in either an objective or subjective fashion. The subjective approach leans toward Weberian principles and treats recreational activities as social actions that are meaningful, or made accountable, to participants in many different ways. Already in this chapter we have cited a number of examples of this approach (Burch, 1969; Etzkorn, 1964) and we will have occasion to return to it again in the next section. Another important feature of Hendee and associates' model is that it stresses a developmental approach; it emphasizes socialization for leisure and the concomitant changes that take place through the life cycle. All too little research on leisure socialization has been carried out, though the papers by Yoesting and Burkhead (1973) and Kelly (1974) are notable exceptions. Both stress the significance of the early family and residential environment to leisure learning but also emphasize that leisure socialization is a lifelong process with experience in one activity invariably breeding more experience. It will be recalled that the studies by Burch (1969) and Hendee (1969a), already discussed, suggested that there is a carry-over from childhood recreational experiences to the adult style. Kelly (1974, p. 192) also argues the important point that "the same activity may have differing social meanings and role relations at different times" and suggests that it would be profitable to explore the hypothesis that leisure is more "*identity-seeking* for youth, more *role-related* for parents, more *work* and *community-related* in later years, and more *interpersonal* or *solitary* in retirement" (author's italics).

On the basis of published research it certainly appears as though the occupational milieu of an individual has an important influence on his or her adult leisure style (Gerstl, 1961; Burdge, 1965; Bishop and Ikeda, 1970), though in recent years research on life-styles, particularly in the United States, has begun to stress the extreme variability and pluralism of emerging life-styles in affluent countries, even among those in the same income or occupational category. These variations are based largely on differences of emphasis with regard to a commitment to either work or leisure (Miller and Sjoberg, 1973; Murphy, 1975) and the relations between the two. On this question Parker (1971), for example, has developed a threefold typology of possible work–leisure relationships as follows: (1) leisure as an extension of work; (2) leisure as neutral from work (neither positive or negative effect of work); and (3) leisure as opposition to work (the compensatory theory). On the basis of current trends Devall (1973) has also suggested that a fourth category, work as an extension of leisure as a central life interest, should be added to the typology (see also Martin and Berry, 1974).

In concluding this section on the social forces influencing recreational choice, I would argue quite strongly along with such writers as Cheek (1971) that the group, variously defined in differing circumstances, is by far the most important unit of analysis in leisure studies. It is from the group that the individual derives his values relating to work and leisure, his preferences for various recreational activities, and his attitudes toward different recreational environments. But our knowledge relating to this area is still rather slight and a great deal more research needs to be carried out, especially on the topics of leisure socialization and activity change. Two approaches would seem to be potentially fruitful: (1) detailed in-depth investigations of the leisure behavior of individuals in the same occupational group at the same stage of the life cycle, but perhaps living in quite different residential environments; and (2) developmental panel studies after the manner of LaPage and Ragain's (1974) eight-year camping project. In the final section of the chapter, which now follows, we turn to a consideration of personality as a variable in recreational motivation and preference.

THE INFLUENCE OF PERSONALITY

On the basis of the discussion to this point the reader could well be forgiven for believing that the author adheres strongly to Skinner's (1971) view that "a person does not act upon the world, the world acts upon him" (p. 211). This is not the case, for it is hoped that the model of man that has emerged is that of a thinking, evaluating, remembering, goal-directed individual who does not simply react to stimuli from the environment in a predetermined mechanistic fashion, but has a fair degree of choice in the stimuli that he/she chooses to select or ignore and also learns from the experiences of others. The view put forward here is certainly not that behavior is altered unconsciously and automatically by consequences, but rather that people can and do play an active role in changing both the social and physical environment that impinges on them. The process is bidirectional rather than unidirectional. Moreover, I would agree with Bandura (1974, p. 866) that "environmental control is overstudied, whereas personal control has been relatively neglected." Perhaps nowhere is this more apparent than in the field of recreation and leisure studies where, with only a few notable exceptions (Driver, 1972; Neulinger, 1974), the psychological element in leisure behavior has been relatively ignored, certainly by comparison with the sociological aspects.

Knopf and Driver (1973) have suggested that recreational behavior is best conceptualized as an interrelated system of five sets of variables as follows: (1) antecedent conditions; (2) user aspirations: (3) intervening variables; (4) user satisfactions; and (5) real benefits.

Antecedent conditions include "all things that give direction and strength to recreation behavioral tendency" (Knopf and Driver, 1973, p. 4). This chapter has dealt with many of the relevant antecedent conditions such as environmental and situational factors, past learning, and maturity (stage of life cycle), but so far has placed little emphasis on the important physiological and personality factors such as relative physical strength or need for variety and change.

An investigation of the literature relating to psychological need-states reveals that human behavior is characterized by two opposite forces: the drive toward stability, or complacency, (Raup, 1925) and the drive toward change and variety. Some psychologists such as Cannon (1932) and Stagner (1951) emphasize the stability-seeking aspects of behavior subsumed under the general label of theories of psychological homeostasis or equilibrium, while others (e.g., Klausner, 1968; Halliday and Fuller, 1974) place somewhat greater emphasis on the human need for risk, change, and variety. The equilibrium model has been inspired by traditional models in physics that assert that when the stable state of a system is disturbed the system will seek to revert to an equilibrium state of minimum tension. When translated into human behavior terms, it is clear that this model—that of man as a conscious stress-seeker—allows that people have a considerable degree of autonomy in their behavior. They are free to make choices; they are free to seek as well as to minimize stress. Klausner (1968, p. 138) has argued that

> equilibrium theory has been reified as a model of conscious motivational systems. . . At times people may act to reduce the tension or stress they are experiencing, but there are also times when they act to increase their tensions.

It is clear that these two perspectives have their sociological equivalents in the familiarity and compensatory theories already outlined above; and it is equally clear that only rarely can a detached objective assessment of a recreational activity enable the researcher to judge adequately whether that activity is fulfilling a tension-reducing or tension-seeking function for an individual. A given activity can be one thing at one time and can fulfill a totally different function on another occasion. The same activity can mean different things to different people. On the basis of empirical research on campers, for example, it is quite clear that very large numbers of residents of extensive urban areas view the activity of camping in a crowded campground as an escape from the city or a return to nature. On this point Clark et al. (1971, p. 148) have suggested that

> it is quite possible that users of modern highly developed campgrounds share the same general values as environment-oriented campers but have different notions of what behavior leads to these values and different standards of judging their attainment.

In a personal communication Colley has suggested that individuals need both tension-reducing and tension-seeking experiences at different times (though to markedly different degrees), and that recreational behavior can be classified in a fourfold way as either ameliorative, enriching, quiescent, or dynamic. She suggests that in quiescent experiences people are essentially being acted upon in either an ameliorative or enriching way (her examples here are seeing something of beauty in nature or attending a concert); and that in dynamic experiences energy is expended. For example, she suggests that motorcycle racing may be ameliorative (healing or compensatory) while rock-climbing or parachute-jumping may be enriching. She stresses that in order to classify an activity one way or another it is essential to discover its meaning for the individual concerned. This involves viewing the activity at a particular time in the context of the antecedent conditions such as the personality of the individual, experience level, work situation, and aspirations.

What at first sight appears to be stress-seeking recreational behavior in particular has been the subject of a considerable amount of behavioral science research in recent years. Indeed, in reviewing the literature relating to motivation in recreation, one could almost be forgiven for thinking that adventurous outdoor activities, broadly defined, are by far the most popular forms of recreation in the world today. Thus, studies have been made of the motivational forces involved in such risk-taking activities as wilderness canoeing (Lucas, 1964; Lime, 1970; Peterson, 1971); backpacking and mountaineering (Stone and Taves, 1956; Hendee et al., 1968; Catton, 1969; Shafer and Mietz, 1969; Stankey, 1973; Scott, 1974); caving (Watson, 1966); surfing (Stone, 1969; Irwin, 1973); sport parachuting (Klausner, 1967; Epstein and Fenz, 1962); and hunting (Hendee and Potter, 1971).

However, while the experiences themselves and the satisfactions derived have sometimes been studied in great detail very little research effort has so far probed the personalities of participants, including such components as general self-concept, anxiety level, hostility, alienation, and the like. In large measure the reason for this relative research neglect undoubtedly lies in the extreme difficulty behavioral scientists have encountered in attempting to operationalize the concept of personality (Hall and Lindzey, 1957; Heritage, 1974), which means that virtually any measurement device that is constructed is open to considerable question and debate. But also, as Neulinger (1974, p. 111) stresses, and as I have been arguing strongly throughout this chapter, "The main problem in trying to predict leisure behavior from personality traits may be that any given activity may fulfill different needs for different people or even the same person at different times." Nevertheless these basic problems have not prevented researchers from utilizing certain favored personality measures and applying them to leisure behavior patterns. Following Lansing (1968), Brok (see Neulinger and Brok, 1974), for example, has been focusing on the significance of the personal effectiveness variable and comparing the leisure and work styles of individuals who believe that they are in control of their own destinies ("internals") with those who feel that they are subject to more powerful external forces beyond their control ("externals"). Preliminary results seem to suggest that "internals viewed leisure as more active, refreshing, better, meaningful, pleasant and satisfying than externals" (p. 169). These results are interesting but we are still left with the problem that these findings state nothing about specific activity preferences. Further, the methodology is based on the assumption that personality is a fixed unalterable variable. One study that did not make this assumption examined the free-time behavior of Canadian Air Force personnel living in two very different physical environments, metropolitan Winnipeg and an isolated Arctic station at Fort Churchill (Farina, 1965). This investigator found evidence to suggest that "a radical change of environment may result in the ascendancy of a different constellation of personality factors in a given individual" (p. 90).

In short, this brings us back full circle to the importance of the social and physical environment in which an individual resides. The physical environment provides him with a wide range of objectively possible recreational alternatives, while a combination of knowledge, learned preferences, personality, and social forces markedly restricts the range of choice that the individual regards as being open to him. Changed economic, health, or social circumstances can have the effect of completely altering what a person regards as possible or feasible in the leisure sphere. This suggests that it would be fruitful for the behavioral scientist interested in the perceptual and motivational aspects of recreation to focus particular attention on changes in leisure behavior that may be found to occur when certain other critical changes take place in an individual's life. The transition from college to first job, from being single to being married, or from residence in a large city to a rural area would provide interesting case-studies and would undoubtedly furnish us with new insights on recreational behavior. In this connection, the work of Wohlwill and Kohn (1973) and Wohlwill (1974) on the important concept of adaptation level is a promising research area and is of relevance to many aspects of leisure research. Do the residents of large metropolitan areas adapt to the relative lack of open space, the noise, and congestion? Do visitors to high-intensity-use state parks and popular wilderness areas simply adapt to the higher crowding levels? All the evidence so far seems to point to the fact that over a period of time they do, though far more research is of course needed to determine at what cost, if any.

REVIEW AND PROSPECTS

Using Figure 1 as an approximate organizational framework for a discussion of the recreating individual, the present chapter has moved through a number of stages. The initial emphasis was on the broad cultural and economic forces shaping a person's leisure needs. The main point made here was that an individual's recreational and/or environmental needs are not given physiological or psychological states but are learned dispositions, subject to considerable potential modification and change through time. Then, in view of the fact that the set of recreational opportunities available in a nation, region, or locality plays such an important part in molding recreational activity patterns and preferences, consideration was given to the importance of the place of residence variable. Subsequently, the influence of various social forces such as family, friends, and work mates was discussed, with particular emphasis being placed on what I consider to be the very important, though little researched area of roles in leisure behavior. Finally, attention was focused on the personality variable. This was treated last, after a discussion of the influence of external social and environmental forces, because of the author's conviction that personality is not fixed and unalterable, but rather is only meaningful when looked at in various social and environmental contexts. The logical extension of this argument is that attempts to relate various personality types to specified activity or environmental preferences (e.g., Sonnenfeld, 1969) are likely to be of somewhat dubious value and, I would argue, of application only with reference to one time and place.

A great deal of research in the general area of environmental perception has now been carried out in the recreation field as can be seen, for example, by the large number of entries in Veal's extensive annotated bibliography on the subject (Veal, 1974). However, I would seriously question the extent to which much of this research has been cumulative and/or comparable in terms of the basic concepts and measures used by the investigators. On this question I find myself in wholehearted agreement with a U.S. Forest Service recreation researcher who despairingly reported:

> For sound research planning, I would gladly swap all the highly "significant" correlation coefficients of the past 10 years for a handful of good case studies that yielded some solid conceptual insights to build on (LaPage, 1971, 192).

At present, the indications are that behavioral scientists are a long way from developing any kind of generalized theory of recreational behavior. Indeed, many would probably argue that in view of the highly complex and intangible nature of the leisure field we are as close to this goal now as we shall ever be. I do not take quite such a pessimistic view, and throughout this chapter have suggested a number of research avenues that I consider would be worth pursuing. In particular, I would advocate that much closer attention be paid to the evolution and development of the individual or group's total leisure profile rather than the researcher concentrating on only one facet of that profile. Leigh's (1971) book, *Young People and Leisure*, an appraisal of the leisure behavior of one segment of the population in a mining town in northern England, provides an excellent example of this approach. Such a research thrust, if widely used for a variety of comparable groups in different residential settings would, I believe, help us to identify the substitute leisure-behavior strategies people employ at different stages of their lives, or as they move from one residential environment to another. Similarly, it would enable us to identify more precisely the behavior people consider appropriate to different roles and to pinpoint in a more precise fashion the influence of family, peers, and work mates. This research emphasis, too, inevitably means that researchers would have to make greater use of relatively unstructured ethnographic techniques than has hitherto been the norm in recreation research. If carried out carefully by skilled researchers, I believe that this approach will begin to furnish the solid conceptual insights many researchers and recreation planners are desperately seeking.

REFERENCES

Albrecht, S. L., DeFleur, M. L., and Warner, L. G. Attitude-behavior relationships. A reexamination of the postulate of contingent consistency. *Pacific Sociological Review*, 1972, *15*, 149–168.

Appleton, I. *Leisure research and policy*. Edinburgh: Scottish Academic Press, 1974.

Bandura, A. Behavior theory and models of man. *American Psychologist*, 1974, *29*, 859–869.

Baumann, D. D. *The recreational use of domestic water-supply reservoirs: Perception and choice*. The University of Chicago, Department of Geography, Research Paper No. 121, 1969.

Bettelheim, B. *The informed heart*. London: Paladin, 1970.

Bexton, W. H., Heron, W., and Scott, T. G. Effects of decreased variation in the se ory environment. *Canadian Journal of Psychology*, 1954, *8*, 70–76.

Bishop, D. W., and Ikeda, M. Status and role factors in the leisure behavior of different occupations. *Sociology and Social Research*, 1970, *54*, 190–209.

Bossard, J. H. S. The concept of family life-cycle. *Journal of American Society of Chartered Underwriters*, 1960, *14*, 308–322.

Brown, P. J. Sentiment changes and recreation participation. *Journal of Leisure Research*, 1970, *2*, 264–268.

Brown, P. J., Dyer, A., and Whaley, R. S. Recreation research—so what? *Journal of Leisure Research*, 1973, *5*, 16–24.

Burch, W. R. The playworld of camping: Research into the social meaning of outdoor recreation. *American Journal of Sociology*, 1965, *70*, 604–612.

Burch, W. R. The social circles of leisure: Competing explanations. *Journal of Leisure Research*, 1969, *1*, 125–147.

Burch, W. R. Recreation preferences as culturally determined phenomena. In B. L. Driver (Ed.), *Elements of outdoor recreation planning*. Ann Arbor: The University of Michigan Press, 1974.

Burch, W. R., and Wenger, W. D. *The social characteristics of participants in three styles of family camping*. USDA Forest Service Research Paper PNW-48, 1967.

Burdge, R. J. *Occupational influences on the use of outdoor recreation*. Unpublished PhD dissertation. Department of Agriculture, Economics and Rural Sociology, The Pennsylvania State University, 1965.

Burdge, R. J., and Field, D. R. Methodological perspectives for the study of outdoor recreation. *Journal of Leisure Research*, 1972, *4*, 63–72.

Burns, T. Leisure in industrial society. In M. Smith, S. Parker, and C. Smith (Eds.), *Leisure and society in Britain*. London: Allen Lane, 1973.

Burton, T. L. Identification of recreation types through cluster analysis. *Society and Leisure*, 1971, *1*, 47–64.

Campbell, F. L. Participant observation in outdoor recreation. *Journal of Leisure Research*, 1970, *2*, 226–235.

Cannon, W. B. *The wisdom of the body*. New York: W. W. Norton, 1932.

Catton, W. R. Motivations of wilderness users. *Pulp and Paper Magazine of Canada*, 1969, December, 121–126.

Cheek, N. H. Toward a theory of not-work. *Pacific Sociological Review*, 1971, *14*, 246–258.

Chubb, M. Recreation use surveys and the ignored majority (mimeo). Paper presented at Michigan State Planning Workshop on User Preference Studies and Demand Analyses. Ann Arbor, Michigan, 1971.

Cicchetti, C. J., Seneca, J. J., and Davidson, P. *The demand and supply of outdoor recreation*. Rutgers University Bureau of Economic Research, New Brunswick, N.J., 1969.

Clark, R., Hendee, J. C., and Campbell, F. L. Values, behavior and conflict in modern camping culture. *Journal of Leisure Research*, 1971, *3*, 143–159.

Clawson, M. Emerging American life style. In M. N. and C. R. Hormachea (Eds.), *Recreation in modern society*. Boston: Holbrook Press, 1972.

Clawson, M., and Knetsch, J. *Economics of outdoor recreation*. Baltimore: Johns Hopkins Press, 1966.

Cosgrove, I., and Jackson, R. *The geography of recreation and leisure*. London: Hutchinson and Co., 1972.

Coughlin, R. E., and Goldstein, K. A. *The public's view of the outdoor environment as interpreted by magazine ad-makers*. Regional Science Research Institute Discussion Paper Series, No. 25, 1968.

Craig, W. Recreational activity patterns in a small negro urban community. *Economic Geography*, 1972, *48*, 107–115.

Crapo, D. and Chubb, M. *Recreation area day-use investigation techniques. Part 1: A study of survey methodology.* Technical report No. 6, Department of Park and Recreation Resources, Michigan State University, 1969.

Devall, W. The development of leisure social worlds. Paper presented at American Sociological Association Convention, New York, 1973.

Driver, B. L. Potential contributions of psychology to recreation resource management. In J. F. Wohlwill and D. H. Carson (Eds.), *Environment and the social sciences: Perspectives and applications.* American Psychological Association, 1972.

Driver, B. L., and Tocher, S. R. Toward a behavioral interpretation of recreational engagements, with implications for planning. In B. L. Driver (Ed.), *Elements of outdoor recreation planning.* Ann Arbor: The University of Michigan Press, 1974.

Dubin, R. Industrial workers' worlds: A study of "central life interests" of industrial workers. *Social Problems*, 1956, *3*, 131–142.

Elson, M. J. *Activity spaces and recreation trip behavior.* Oxford Working Papers in Planning Education and Research, Oxford Polytechnic, Department of Town Planning. Paper No. 19, 1974.

Ennis, P. H. Leisure in the suburbs: Research prolegomenon. In W. M. Dobriner (Ed.), *The suburban community.* New York: G. P. Putman, 1958.

Epstein, S., and Fenz, W. D. Theory and experiment on the measurement of approach–avoidance conflict. *Journal of Abnormal and Social Psychology*, 1962, *64*, 97–112.

Etzkorn, K. P. Leisure and camping—the social meaning of a form of outdoor recreation. *Sociology and Social research*, 1964, *49*, 76–89.

Farina, A. J. O. *A Study of the relationship between personality factors and patterns of free-time behavior.* PhD dissertation, Washington University, George Brown School of Social Work, 1965.

Farina, J. Toward a philosophy of leisure. *Convergence, An International Journal of Adult Education*, 1969, *2*, 14–16.

Gans, H. Outdoor recreation and mental health. In ORRRC Study Report 22, *Trends in American living and outdoor recreation.* Washington, D.C.: U.S. Government Printer, 1962.

Gerstl, J. F. Leisure, taste and occupational milieu. *Social Problems*, 1961, *19*, 56–68.

Green, A. *Recreation, leisure and politics.* New York: McGraw-Hill, 1964.

Hall, C. S., and Lindzey, G. *Theories of personality.* New York: John Wiley, 1957.

Halliday, J., and Fuller, P. *The psychology of gambling.* London: Allen Lane, 1974.

Hauser, P. M. Demographic and ecological changes as factors in outdoor recreation. In ORRRC Study Report 22. *Trends in American Living and outdoor recreation.* Washington , D.C.: U.S. Government Printer, 1962.

Havighurst, R. J. Education, social mobility and social change in four societies. *International Review of Education*, 1958, *4*, 167–183.

Hecock, R. D., and Rooney, J. *The impact of a major new reservoir on recreation behavior.* Department of Geography, Oklahoma State University, 1972.

Hemmens, G. C. Analysis and simulation of urban activity patterns. *Socio-Economic Planning Sciences*, 1970, *4*, 53–66.

Hendee, J. C. Rural–urban differences reflected in outdoor recreation participation. *Journal of Leisure Research*, 1969a, *1*, 333–341.

Hendee, J. C. Appreciative versus consumptive uses of wildlife refuges: Studies of who gets what and trends in use. *Transactions of Thirty-Fourth North American Wildlife and Natural Resources Conference,* Washington, D.C.: Wildlife Management Institute, 1969b.

Hendee, J. C. Sociology and applied leisure research. *Pacific Sociological Review,* 1971, *14,* 360–368.

Hendee, J. C., and Potter, D. R. Human behavior and wildlife management: Needed research. *Transactions of the Thirty-Sixth North American Wildlife and Natural Resources Conference.* Washington, D.C.: Wildlife Management Institute, 1971.

Hendee, J. C., Catton, W. R., Jr., Marlow, L. D., and Brockman, C. F. *Wilderness users in the Pacific Northwest—Their characteristics, values, and management preferences.* USDA Forest Service Research Paper PNW-61, Pacific Northwest Forest and Range Experiment Station, Portland, Oregon, 1968.

Hendee, J. C., Gale, R. P., and Catton, W. R., Jr. A typology of outdoor recreation activity preferences, *Journal of Environmental Education,* 1971, *3,* 28–34.

Hendee, J. C., and Burdge, R. J. The substitutability concept: Implications for recreation research and management. *Journal of Leisure Research,* 1974, *6,* 157–162.

Hendricks, J. Leisure participation as influenced by urban residence patterns. *Sociology and Social Research,* 1971, *55,* 414–428.

Heritage, J. Assessing people. In N. Armistead (Ed.), *Reconstructing social psychology.* Harmondsworth: Penguin, 1974.

Hollander, O. *American and soviet society.* Englewood Cliffs, N.J.: Prentice-Hall, 1969.

Irwin, J. Surfing: The natural history of an urban scene. *Urban Life and Culture,* 1973, *2,* 131–160.

Kaplan, M. Leisure as an issue for the future. *Futures,* 1968, *1,* 91–99.

Kaplan, M. and Bosserman, P. (Eds.), *Technology, human values and leisure,* N.Y.: Abingdon, 1971.

Kaplan, M., and Lazarsfeld, P. The mass media and man's orientation to nature. In ORRRC *Study Report 22, Trends in American living and outdoor recreation.* Washington, D.C.: U.S. Government Printer, 1962.

Kelly, J. R. Socialization toward leisure; a developmental approach. *Journal of Leisure Research,* 1974, *6,* 181–193.

Kerr, W. *The decline of pleasure.* New York: Simon and Schuster, 1965.

Klausner, S. Z. Fear and enthusiasm in sport parachuting. In J. A. Knight and R. Slovenko (Eds.), *Motivations in play, games and sports.* Springfield, Ill.: Charles C. Thomas, 1967.

Klausner, S. Z. *Why man takes chances. Studies in stress-seeking.* New york: Doubleday-Anchor, 1968.

Knetsch, J. L. A design for assessing outdoor recreation demands in Canada (Mimeo), 1067.

Knopf, R. C., and Driver, B. L. A problem-solving approach to recreation behavior (mimeo). Paper presented at Seventy-seventh annual meeting, Michigan Academy of Science, Arts and Letters, Ann Arbor, 1973.

Knopp, T. B. Environmental determinants of recreation behavior. *Journal of Leisure Research,* 1972, *4,* 129–138.

Lansing, J. B. The effects of migration and personal effectiveness on long distance travel. *Transportation Research,* 1968, *2,* 329–338.

LaPage, W. F. Cultural "fogweed" and outdoor recreation research. In *Recreation Symposium Proceedings.* Northeastern Forest Experiment Station, USDA, Upper Darby, Pa., 1971.

LaPage, W. F. The outdoor recreation culture—implications for commercial development (mimeo). Paper presented at meeting of Society for Range Management, Washington, D.C., 1972.

LaPage, W. F., and Ragain, D. P. Family camping trends—an eight-year panel study. *Journal of Leisure Research.* 1974, *6*, 101–112.

Lee, R. G. The social definition of outdoor recreational places. In W. R. Burch, N. H. Cheek, and L. Taylor (Eds.), *Social behavior, natural resources, and the environment.* New York: Harper and Row, 1972.

Leigh, J. *Young people and leisure.* London: Routledge and Kegan Paul, 1971.

Lime, D. Research for determining use capacities of the Boundary Waters Canoe Area. *Naturalist,* 1970, *21*, 8–13.

Linder, S. B. *The harried leisure class.* New York: Columbia University Press, 1970.

Lucas, R. C. Wilderness perception and use: The example of the Boundary Waters Canoe Area. *Natural Resources Journal,* 1964, *3*, 394–411.

McNeal, J. U. (Ed.), *Dimensions of consumer behavior.* New York: Appleton-Century-Crofts, 1965.

Martin, T. W., and Berry, K. J. Competitive sport in post-industrial society: The case of the motocross racer. *Journal of Popular Culture,* 1974, *8*, 107–120.

Maslow, A. B. *Motivation and personality.* New York: Harper and Row, 1954.

Maw, R., and Cosgrove, D. *Assessment of demand for recreation—modelling approach.* Working paper 2, Leisure Model Unit, Built Environment Research Group, Polytechnic of Central London, 1972.

Mercer, D. C. The role of perception in the recreation experience. *Journal of Leisure Research,* 1971, *3*, 261–276.

Mercer, D. C. The concept of recreational "need." *Journal of Leisure Research,* 1973, *5*, 37–50.

Mercer, D. C. Recreation for those at a social disadvantage: The case of women. In *Leisure: A new perspective.* Canberra: Australian Government Printer, 1974.

Mercer, D. C. Perception and outdoor recreation. In P. Lavery (Ed.), *Recreational Geography.* Newton Abbott: David and Charles, 1975.

Miller, P. J., and Sjoberg, G. Urban middle class life styles in transition. *Journal of Applied Behavioral Science,* 1973, *9*, 144–161.

Mueller, E., and Gurin, G. *Participation in outdoor recreation: Factors affecting demand among American adults.* ORRRC Study Report 20. Washington, D.C.: U.S. Government Printer, 1962.

Murphy, J. F. *Concepts of leisure.* Englewood Cliffs, N.J.: Prentice-Hall, 1974.

Murphy, J. F. *Recreation and leisure service: A humanistic perspective.* Dubuque, Iowa: William C. Brown, 1975.

Murphy, P. E., and Rosenblood, L. Tourism: An exercise in spatial search. *Canadian Geography,* 1974, *18*, 201–210.

Neulinger, J. *The psychology of leisure.* Springfield, Ill.: Charles C. Thomas, 1974.

Neulinger, J., and Brok, A. J. Reflections on the 1973 American Psychological Association Symposium on Leisure. *Journal of Leisure Research,* 1974, *6*, 168–171.

Outdoor Recreation Resources Review Commission. Study Report 3. *Wilderness and recreation: A report on resources, values and problems.* Washington, D.C.: U.S. Government Printer, 1962a.

Outdoor Recreation Resources Review Commission. Study Report 19. *National Recreation Survey.* Washington, D.C.: U.S. Government Printer, 1962b.

Outdoor Recreation Resources Review Commission. Study Report 20. *Participation in outdoor recreation: Factors affecting demand among American adults.* Washington, D.C.: U.S. Government Printer, 1962c.

Parker, S. *The future of work and leisure.* New York, Praeger, 1971.

Peterson, G. L. *Motivations, perceptions, satisfactions and environmental dispositions of Boundary Waters Canoe Area users and managers.* Northwestern University, Department of Civil Engineering, Evanston, Ill., 1971.

Proctor, C. Dependence of recreation participation on background characteristics of sample persons in the September 1960 National Recreation Survey. *ORRRC Study Report 19.* Washington, D.C.: U.S. Government Printer, 1962.

Raup, R. B. *Complacency: The foundation of behavior.* New York: Macmillan, 1925.

Rodgers, H. B. Problems and progress in recreation research: A review of some recent work. *Urban Studies,* 1972, *9,* 223–228.

Schatzman, L., and Strauss, A. *Field research: Strategies for a natural sociology.* Englewood Cliffs: Prentice-Hall, 1973.

Scott, N. R. Toward a psychology of wilderness experience. *Natural Resources Journal,* 1974, *14,* 231–237.

Shafer, E., and Mietz, J. Aesthetic and emotional experiences rate high with north-east wilderness hikers. *Environment and Behavior,* 1969, *1,* 186–197.

Skinner, B. F. *Beyond freedom and dignity.* New York: Knopf, 1971.

Sonnenfeld, J. Variable values in space landscape: An inquiry into the nature of environmental necessity. *The Journal of Social Issues,* 1966, *22,* 71–82.

Sonnenfeld, J. Personality and behavior in environment. *Proceedings of the Association of American Geographers,* 1969, *1,* 136–140.

Stagner, R. Homeostasis as a unifying concept in personality theory. *Psychological Review,* 1951, *58,* 5–17.

Stankey, G. H. *Visitor perception of wilderness recreation carrying capacity. USDA Forest Service Research Paper INT-142, 1973.*

Steele, P. D., and Zurcher, L. A. Leisure sports as "ephemeral roles." *Pacific Sociological Review,* 1973, *16,* 345–356.

Stone, G., and Taves, M. Research into the human element in wilderness use. *Proceedings of the Society of American Foresters,* 1956, 26–32.

Stone, R. E. *Meanings found in the acts of surfing and skiing.* PhD dissertation, University of Southern California, 1969.

Tatham, R. L., and Dornoff, R. J. Market segmentation for outdoor recreation. *Journal of Leisure Research,* 1971, *3*(5), 5–16.

Turnbull, C. M. *The mountain people.* London: Pan, 1974.

Veal, A. J. *Environmental perception and recreation: A review and annotated bibliography.* Research memorandum 39, Centre for Urban and Regional Studies, University of Birmingham, 1974.

Watson, R. A. Notes on the philosophy of caving. *National Speleological Society News,* 1966, *24,* 54–58.

West, P. C., and Merriam, L. C. Outdoor recreation and family cohesiveness. *Journal of Leisure Research,* 1970, *2,* 251–259.

Wohlwill, J. F. Human adaptation to levels of environmental stimulation. *Human Ecology,* 1974, *2,* 127–147.

Wohlwill, J. F., and Kohn, I. The environment as experienced by the migrant. *Representative Research in Social Psychology,* 1973, *4,* 135–164.

Wolfe, R. I. Vacation homes and the gravity model. *Ekistics,* 1970, 352–353.

Yoesting, D. R., and Burkhead, D. L. Significance of childhood recreation experience on adult leisure behavior: An exploratory analysis. *Journal of Leisure Research,* 1973, *5,* 25–36.

5

Work Environments

H. McILVAINE PARSONS

How do environments influence human behavior in the world of work? More particularly, how do work environments affect the workers? This chapter will examine two environments that can be designed to influence worker behavior: industrial plants and business offices.

EFFECTS ON WORKERS: A FRAMEWORK

To examine how work environments affect workers, I shall adapt a framework for analyzing how constructed environments in general influence the behavior of those who use them (Parsons, 1974a). It gives precedence to what happens to the worker, the dependent variable.

Job Activities. These are what a worker does on the job, the component actions in assembling an automobile or maintaining cleri cal files (or in the home, cooking a meal), measured in working effectiveness, productivity, or quality of output (e.g., errors).

Feelings. Workers express satisfaction or dissatisfaction, enjoyment or distaste, comfort or distress, which stem in part from physical surroundings. They can be probed through interviews and questionnaires and spontaneous suggestions or grievances, as well as physiological correlates.

Manipulation. Some workers can manipulate their work environ-

H. McILVAINE PARSONS · Institute for Behavioral Research, Inc., Silver Spring, Maryland.

ments by buying or selling them, or through decisions or preferences about their design. Others do so by changing jobs or locations, taking sick leave, or coming to work late, as indicated in turnover and absenteeism data.

Health and Safety. Work environments can have toxicological effects. They can be designed to reduce accidents, or their maldesign may make these more probable. Records of medical treatment can indicate frequency and severity.

Social Interaction. Although this aspect of behavior (and the others that follow) have less bearing on the design of *work* environments, social interactions do occur in work environments, which can be designed either to facilitate or hinder them.

Motivation. Behavior is motivated by its consequences and by the potentiators that make these effective. They strengthen, maintain, or weaken our habits, including work habits, and some get designed into the physical environment, perhaps inadvertently. If a work situation is disagreeable, effortful, or uncomfortable, the worker will tend to avoid, leave, or terminate it.

Locomotion. Walking between job activities or reaching them from the outside can have importance for the worker, especially if he or she is disabled, though the designer of the work environment may pay it little heed.

Learning. People who use an environment have to learn to get around in it. Other effects akin to learning occur. Novelty wears off. Workers adapt to disagreeable conditions. Places acquire meanings for those using them. Reactions to a new or changed work setting may depend on its resemblance to the old. Learning affects perception, which depends on an individual's experience in an environment as well as its actualities.

Perception. An operational definition of environmental perception is simply talking or writing or thinking about an environment and its attributes, or sketching or imagining it. Obviously, it involves memory. Little research on environmental perception has been directed at work environments.

WORK ENVIRONMENTS: A SECOND FRAMEWORK

What about the environment? In this chapter environment means *physical* surroundings that *can* be designed or arranged to influence work behavior. That leaves out people and the total milieu or culture, important as these are. It also leaves out the physical environment that is not, and by and large cannot, be designed or arranged to have an effect on the workers in it.

We must also consider a setting's size or scope. This chapter will exclude on the one hand constructed environments that the worker cannot sense directly (e.g., the community) and on the other various focal points of attention, i.e., individual signals, displays, or manipulanda (such as a warning light, a dial face, or a switch). Just where to draw the line between a setting and its components is a question to which no answers are found in the literature of environmental design and environmental psychology.

There are various ways of examining a work environment—or any constructed environment, for that matter—and of altering its design to affect workers. Eight categories comprise a framework that can be considered in connection with the framework of user reactions.

Entire Setting. A setting can be viewed in its entirety. When a worker changes jobs, he or she may move to a work environment that differs in every respect though in varying degree.

Resources. These are the physical equipment a worker uses: an assembly line or filing cabinets; in the home, a stove. First, required resources must be *provided*: no drilling machine, no drilling; at home, no stove, no cooking. Second, they must be suitably *located* for the individual worker; he or she should have easy access to the filing cabinet. Third, they must be properly *designed* for human use: A worker should be able to read gauges without error. Resources are particularly important to a worker's job activities (the first category in the user framework) and his health and safety.

Spatial Arrangement. Walls (exterior and interior), partitions, ceilings, stairs, openings (doors, windows), furniture, and equipment determine through their locations and dimensions the sizes and interrelationships of spaces (areas, rooms, closets, halls, and corridors). In turn these establish where workers work, their paths from place to place, and their density, proximities, and visual or auditory access to each other (including privacy). Spatial arrangements especially influence the user categories of locomotion, social interaction, and learning.

Ambient Conditions. In industrial plants and offices, ambient conditions include illumination, noise, vibration, temperature, humidity, atmospheric composition, and odors (and in other work environments they include acceleration, reduced gravity, radiation, and decreased or increased barometric pressure). Some conditions can be designed into the setting (e.g., illumination); some can be controlled by means of design (e.g., temperature); and some result from, and thus are subject to, the design of parts of the work environment (e.g., noise and vibration). Suboptimal ambient conditions affect workers' job activities, feelings, motivation, and health and safety.

Communication. Through their design, constructed environments and their components tell their users what to do, and work environments are no exception; a work bench in itself indicates one kind of required activity; a desk, another; a sink, still something else. Less obviously, work environments can incorporate alerting, timing, warning, instructional, and feedback signals or indicators—and even exhortations (e.g., Think or Quiet). Special codes may be installed, especially to guide workers from place to place. Communication between persons may require telephone and public address systems, or acoustical treatment of spaces. Communication thus affects job activities, locomotion, and social interaction.

Consequation and Potentiation. Consequences can be built into designed settings as rewards and punishments, and sometimes are— but perhaps more often unintentionally. How might pleasant surroundings be made contingent on coming to work, or on working? To what extent do physical arrangements that require undue exertion or discomfort result in reduced performance? Do some disagreeable work conditions potentiate a worker's tendencies to get away from them? The motivational effects of the physical work environment are often obscured by the competing motivational effects of the social and organizational environments.

Protection. All structures, work settings included, are intended to protect those within from external events and conditions that might damage their health (e.g., from rain and snow, intense heat or freezing weather, and even harmful human agents). Within structures are hazards. Work settings include many opportunities for accidents that proper design could prevent, or whose effects it could minimize. It is also possible to protect workers against extreme ambient conditions that arise within the structure. Thus, this category of environmental design is most closely related to the effects of that design on users' health and safety.

Appearance. What a work environment looks like may have an effect on the workers in it, even though its visual appearance is more likely to result from its effects on the designer. The appearance of exteriors, interiors, and environs has been a major architectural concern, with the designer rather than the user as judge of what the appearance should be. Although not much interest has been expressed in workers' reactions to the appearance of their work settings, presumably it can affect both their feelings and perception.

These eight factors are not all wholly independent of each other, just as those in the first framework also overlap. The purpose of each set of categories is simply to make the analyses of user reactions and work environments more manageable.

SETTINGS

Now we shall match pertinent categories in these two frameworks against each other in the two work settings.

INDUSTRIAL PLANTS

Job Activities

The job activities of workers are influenced the most by resources, spatial arrangement, and ambient conditions. The effects these have on performance have been investigated by industrial engineers (e.g., under the label of time and motion study) and by engineering psychologists and others in the field of human factors (or ergonomics, the European name for this field). Until relatively recently, industrial engineers concentrated on manufacturing and human factors people focused on military systems.

Resources. This environmental category means *providing* machinery for workers to use, *locating* it, and *designing* it so that they can use it effectively

Some equipment is not provided to a worker in an industrial plant because it has been superseded by automatic devices. When this has happened, job activities have changed as a result, especially where a computer gets signals directly from what is being processed and directly activates the processing controls (e.g., valves or switches). The extent to which the equipment environment is automated affects what the worker does. With automation, the operator monitors displays instead of manipulating control devices. Maintenance workers have to inspect and repair sensing and control circuits as well as the processing equipment itself. As a result of precomputer automation, various devices lift, carry, and emplace materials that earlier had to be handled manually.

In manufacturing operations, automated or otherwise, there may be some job-related activities for which no resource is provided because the particular activity was overlooked, deemed trivial, or excluded from the job. Though it might be possible for a worker to accomplish a task while seated, there may be no place to sit, even for a moment of rest. Temporary storage for some tool may be lacking. Of more consequence, if the job was designed to include only a certain number of operations, the equipment for other operations will not be present and the job cannot be enriched by adding these other operations. More will be said about job enrichment in connection with spatial arrangement.

The location of machinery in relation to the individual industrial worker was analyzed through time and motion study (Gilbreth, 1911; Gilbreth and Gilbreth, 1923). More recently, Barnes (1949) noted that "tools, material, and controls should be located close in and directly in front of the operator." Normal and maximum working areas are specified for each hand of a seated or standing operator. To reduce the time required for manual tasks, such as assembly, the items to be assembled are placed to minimize arm motion and use both hands at the same time. Just a fraction of a second saved in each operation could amount to a substantial economy for an entire department during a month. "In the continuous or progressive type of manufacturing, machines, process apparatus, and equipment should be arranged so as to require the least possible movement on the part of the operator," Barnes (1949) added, and "materials and tools should be located to permit the best sequence of motions."

More recently, human factors engineers and psychologists have been concerned with the location of controls and displays in relation to the operator of equipment in man–machine systems, such as an aircraft pilot with a cockpit full of instruments or an air traffic controller at a console. Here, tasks demand not so much the highest possible speed of repetitive hand movements as they do the ability to reach one of many switches, precisely manipulate a stick or knob, and accurately discriminate various kinds of dials and counters and information on cathode ray tubes and other displays. As control consoles are increasingly introduced into some kinds of industrial plants (e.g., for chemicals processing and energy production and distribution), many of the data on design requirements for such work environments become applicable to industry.

Design requirements for consoles (McCormick, 1970; Van Cott and Kinkade, 1972) include the arrangement of displays and controls, visual and manual access to them, and human body support. Especially for large wrap-around consoles, the designer must determine how displays (and associated controls) should be placed in relation to each other. Guiding principles for panel arrangement are sequence of use, frequency of use, importance of use, and functional grouping.

Rarely can all of these be put in effect at the same time. For example, an emergency switch may be infrequently used, but it should be highly accessible to the operator because of its importance.

Because workers differ so much in their body measurements, console design—and workplace design in general—should take such anthropometric differences into account lest it hamper the performance of some individuals. Anthropometric data assembled by Damon, Stoudt, and McFarland (1966) are applicable to various manufacturing tasks besides operating a console, for example, in making certain that good postures can be maintained at seated or standing tasks. Ayoub (1973) has established the range of heights of work surfaces for fine assembly work, mechanical assembly work, light assembly work or writing, and coarse manual work, such as packaging. Anthropometric data are also applicable to the design of chairs, which are discussed later in the section on office environments.

Although automatic devices have replaced most strenuous manual labor, what remains warrants studying the biomechanical aspects of the industrial environment. The strength a worker can exert on a handwheel depends on where this is situated in relation to the worker. Individual differences in strength are of consequence (Damon et al., 1966). The work setting may require muscular effort in lifting, pushing, pulling, torque, and carrying.

In addition to the *provision* of resources (equipment) and their *location* in relation to the individual worker, we must consider their *design*, although *individual* controls and displays, the subject of most human engineering, are excluded from this analysis. Ensembles of these often make up "surrounds" of individual workers. Examples include control rooms in refineries, power plants, and power or water distribution centers. These contain large wall displays of pipelines or power lines with intervention points, as well as status indicators of various types. Their design coding, discriminability between elements, indications of options for decision making, and display of actions to take in emergencies—can be of considerable significance in determining the performance of workers who staff these nerve centers.

Spatial Arrangement. The overall arrangement of machinery in a factory has a major effect on the job activities of workers because it determines for many the designs of their jobs and the equipment at which they work. Thus, the physical work environment can limit or expand the scope of job activities, particularly for assembly operations in their several forms (Muther, 1956):

1. *Fixed location:* Materials, tools, workers, and machinery are concentrated in one place; everything is brought there, and a small

team or perhaps, in some cases, an individual worker accomplishes the entire assembly.

2. *Production line:* The material moves past a series of workers, each of whom performs some assembly task or a small set of tasks. If the parts to be assembled are carried on a belt, this moves at some fixed pace, and tasks are performed in an established sequence.

3. *Function layout:* All operations of a particular type of process, such as welding, drilling, or painting, are accomplished in the same area.

Various system and subsystem criteria have been applied in assessing the relative advantages of these work environments from a management viewpoint: time, cost, flexibility, machine breakdown, machine investment, and workers' proficiency, incentives, accountability, supervision, absenteeism, training, and availability. One type of layout may be preferred to another according to the nature of the materials, machinery, workers, movement, time and service factors, and features of the building. Although the effects on the workers and their reactions to the kind of layout should be important both to management and to the workers, their reactions have had little impact, notably when the second type of layout, the mechanized assembly line, was introduced in the early 1900s as the "scientific management" approach to assembly in manufacturing advocated by Frederick W. Taylor.

An assembly line influences job performance directly in two ways that are pertinent to overall production. First, a worker must perform at the pace set by the line; he cannot set his own pace. This disregards differences in work rates between workers, differences within workers during the day or from day to day, and possible advantages of self-pacing. Second, errors may still occur, often intentionally, as recent literature attests (e.g., Lawler, 1973; Terkel, 1974). Although the *amount* of these reflects the workers' motivation and will be discussed under that topic, the *kinds* of errors result, at least in part, from the nature of the work environment. Presumably, the kinds of errors in fixed-location layouts would be different.

The fixed-location approach permits—indeed, may require—what has been called job enrichment. This means enlarging or expanding a job horizontally or vertically or both by giving it a variety of tasks, at approximately the same skill level or at different skill levels. Lawler (1969) examined 10 studies of job enrichment. Production increased in quality in all, in quantity in six. Some kinds of workers react differently to job enrichment than do others (Blood and Hulin, 1967; Turner, 1965). Job enrichment programs for assembly operations (the type of job activities to which they are appropriate) have been established, for example, in Motorola's Pageboy II receiver plant, Corning Glass's hotplate and biomedical instrument manufacturing operations, and Maytag's assembly of automatic washer control panels, as well as some European automotive assembly plants. Probably the first fixed-location assembly operations to be studied intensively were those in the Relay Assembly Test Room experiment at the Hawthorne Works of the Western Electric Co. (Roethlisberger and Dickson, 1939). What should be kept in mind is that the layout of machinery must be changed to make job enrichment possible. Shifting from assembly line to fixed-location assembly operations to enrich jobs constitutes a substantial alteration in an industrial plant's physical work environment.

Ambient Conditions. In industrial plants, the ambient conditions that can affect workers' job activities (performance) include sound, vibration, temperature, and illumination.

To what extent does sound in the form of noise degrade performance? Injury to hearing, i.e., hearing loss, clearly hampers performance that depends on auditory communication. Injurious noise will be considered later under health and safety. What about noninjurious noise? Kryter (1970) pointed to inconsistencies in laboratory results, and it has been shown that the effects of noise on performance depend on the nature of the task (Theologus, Wheaton, and Fleishman, 1974), as well as the intensity level (dB), frequency spectrum, and extent and nature of intermittency or predictability. Further, individuals differ in their reactions, and listeners adapt.

The nature of the task seems critical. As much research has shown, noise at even high intensities fails to affect production rate or incidence of errors in most tasks, and may even improve a person's performance. It is erroneous to assume that loud noise will *generally* interfere with what one does, in a factory or elsewhere. But noise can mask what another person says or conceal an auditory warning signal, so one either fails to hear the person or signal or discriminates him or it incorrectly. Such masking occurs frequently in industrial plants, where high-intensity noise is widespread. Failure to hear a warning signal may result in an accident or production mishap. In noise, the intensity of speech must be raised to make it detectable and comprehensible. People shout and yell, or resort to pointing or other visual signals; these activities have to be standardized and practiced to be interpreted correctly.

In the study by Theologus and associates, subjects took longer to press a button in response to a signal in a reaction time task during randomly intermittent noise than in quiet, but these conditions failed to produce any difference in a tracking task (or in a second session of the reaction time task). Carroll (1973) also has shown that loud sound (tones) slow simple reaction times. But most recent laboratory research on the effects of noise on performance has been addressed to tasks that require subjects to discriminate *between* stimuli. These are sometimes called complex monitoring or vigilance tasks. For example, in intense intermittent noise subjects were able to discriminate one letter from a number of letters less often than in quiet (Warner and Heimstra, 1971). Some investigators have required subjects to watch dials and report when the needle on one of them was deflected, or to react when one of five lamps in a circular pattern lit up. Although Kryter (1970) has expressed doubts about some of the earlier studies, it has been suggested that performance on such tasks suffers during noise because the tasks demand division of attention. Simulations of proofreading have also been tested in noise. If further research demonstrates consistency in results, it may be justifiable to extrapolate the laboratory data to the factory. Inspection activities somewhat resemble complex monitoring tasks. An inspector may have to detect a defective item among many satisfactory items. It is possible that factory noise may degrade his performance. On the other hand, the simple reaction time data are probably not applicable, partly because they seem to adapt out, partly because the time differences are too small to affect what workers have to do in industrial plants.

According to laboratory research, randomly intermittent noise that is "unexpected" or "unpredictable" seems more likely to have an effect than continuous or periodic noise. Both kinds can be encountered in inspection operations in factories. It has been found that a person makes errors during proofreading *after* randomly intermittent noise has ceased, whereas continuous noise did not have this aftereffect (Glass and Singer, 1972). The investigators found no adverse effect of noise *during* tests of addition and number comparison. Weinstein (1974) found that *during* irregular noise no more typographical errors and misspellings were missed than during quiet, although in noise proofreaders detected fewer errors in grammar and missing or wrong words.

In this last study, the noise consisted of a recording of a teletype machine. Generally, laboratory studies have used synthetic white noise. Neither typifies the noises in manufacturing. Broadbent and Little (1960) investigated factory operations directly in a plant that produced motion picture film. They concluded that although noise reduction failed to affect the rate of work, it led to fewer errors.

However, Kryter (1970) suggested the latter result may have been due to a reduction in sound interference. Further on-site studies will come about as industry complies with governmental regulations to reduce noise levels to protect against hearing loss.

The direct effect of sound on job activities, when it changes these at all, is not necessarily adverse. Sound, including noise, could benefit performance. Depending on the occasion and nature of the noise, it can alert or arouse a worker. Studies by Smith (1951) and Corso (1952) indicated greater output (and associated errors) in clerical and psychomotor tests during noise than in quiet. Some research has supported the notion that noise may reduce errors in a monotonous task. It has also been found, as reviewed by Kryter (1970), that sleep-deprived subjects "generally performed better in 100 dB and 90 dB random noise than in the quiet, particularly in the later parts of the test sessions." Further, if noise constitutes a signal to do something that otherwise might be neglected, such activity might be more likely to occur. Although signals are usually discrete and would call for occasional, brief noise rather than repetitive or continuous noise, it seems possible that continuing noise might become associated with some activity as a "setting" kind of signal or discriminative stimulus for maintaining that activity.

Sound of a certain type, music, has been intentionally introduced into factories to improve performance. As will be noted later, workers for the most part have liked what they heard, and employers have perhaps assumed that as a result they worked harder. Although no evidence exists that if someone enjoys his or her environment or some aspect of it, more or better output will result, industrial managements have introduced music widely. Before the development of public address systems over which canned music could be played, factories employed live musicians: bands, orchestras, and choruses by the hundreds. Popular articles supported this custom, and relatively little heed has been paid to the limited amount of careful research that examined its efficacy. However, McGehee and Gardner (1949) reported that music had no effect on the productivity of a group of employees in a plant manufacturing rugs, although most of these said they wanted the music to continue and thought it made them more productive. Some years later similar results were found in a study of the effects of music in a skateboard factory (Newman, Hunt, and Rhodes, 1966). It had no influence on number of units produced or percentage of rejects, though the employees liked it and wanted its continuation. Although it has been suggested that music would be most likely to affect low-skill monotonous work, this was the kind of work these employees were doing. Smith (1961) could discover no effect of music on key-punch operating. On the other hand, Wyatt and Langdon (1937) reported increases in output up to 6% when music was played during a repetitive assembly task. In other research reviewed by Uhrbrock (1961), methodology has been questionable and validity doubtful.

Another ambient condition in industrial plants is vibration. Machines vibrate. So do vehicles and tools. Workers on surfaces that vibrate or who hold equipment that vibrates also vibrate. As a result, vision is blurred, acuity diminished, and fixation made more difficult. Workers find it harder to read numbers or letters on dials. The difficulty varies according to the distance of the display from the eye and the amplitude and frequency of vibration, as well as what vibrates: person, display, or both. Manual performance is also impaired. Tasks that "require steadiness or precision of muscular control are likely to show decrements from vibration" (Grether, 1971), as are tracking tasks, which call for manual movements and visual following. On the other hand, reaction time, monitoring, and pattern recognition are not affected by vibration, although they involve central processes.

Research on the effects of vibration on performance has been conducted both in laboratories and in vehicles (both aircraft and wheeled) but to little or no extent in factories. Yet, there exists an abundance of vibration in some industrial plants. In extrapolating from available research data, it would be expected that vibration in factories primarily affects tasks requiring manual dexterity and good visual acuity, such as certain assembly tasks (e.g., electronic equipment), inspection, and precise settings of knobs or other controls.

Heat and cold occur in some industrial plants at levels that can impair work performance. In other words, temperature affects job activities. Ambient temperature within a plant might be too warm if the external temperature was high and the air conditioning, if any, failed to provide adequate cooling, or if machinery or processing operations, e.g., smelting, generated considerable heat that air conditioning or ventilation could not or did not counteract. (Radiant heat is not dissipated by air conditioning.) The internal temperature might be too cool if the external temperature was low and the heating system, if any, failed to provide adequate heat, or if certain plants or locations, e.g., cold storage, required a low ambient temperature. According to Murrell (1969):

> In many industrial processes, men are subjected to radiant heat from furnaces or similar sources; this can rarely occur without in some way affecting the efficiency and output of the operatives. In some instances the

effect may be quite obvious since men have to take rest away from the heat
in order to cool off. In other instances the effect may not be so obvious
since the men's cooling off period may be disguised by breaks in the
process.

A special effect occurs in cold, a loss of manual dexterity; if
workers are protected by gloves, these and the devices they handle
must be designed so the latter can be manipulated properly. Temperature problems are particularly likely to occur, of course, in arctic or
tropical climates.

The relatively meager research conducted within industrial locations on the effects of temperature on job activities has been summarized by Murrell (1969). For example, loading operations in a coal mine
were reported to be 41% less effective above 28°C (82°F) than at 19°C
(66°F). With higher temperatures workers took more frequent rests. In
addition, Huntington (1924) reported that the productivity of 550
pieceworkers in three factories in Connecticut in 1910–1913 was higher
in the months of June, October, and November than in other months.
Most research has come either from laboratories or experiments in hot
or cold climates and the results have to be extrapolated to the
industrial scene, taking into account the various kinds of job activities
there.

Heavy work generates considerable body heat, which is lost more
slowly if a person is in a hot and humid rather than cool and dry
environment. In one study cited by Poulton (1970), precision in a
fairly strenuous task deteriorated when effective temperature rose to
79°F (26°C) with 80% relative humidity and to 84°F (29°C) with 20%
relative humidity. Related research showed deterioration in weight
lifting, the extent and temperature levels at which this developed
varying according to incentives and individual difference. In much
research changes in performance are related to changes in body (e.g.,
skin or rectal) temperature (which in turn is affected by ambient
temperature). It has been reported that an increase in body temperature resulted in more errors in a calculation task but, with sufficient
increase, in greater vigilance due to arousal (Poulton, 1970). Possibly
the latter result could be applied to factory inspection tasks, but it
would be prudent to get more data. Most heat research has been
concerned with metabolic cost, physiological reactions, tolerance,
reported discomfort, and acclimatization, rather than performance; to
be sure, these are related to performance, and "heat stress" will
degrade, even terminate, job activities. According to Berenson and
Robertson (1973), "performance begins to deteriorate in any given
condition at about 75% of the physiological tolerance limit" for heat

stress, and additional recovery time is needed if this is exceeded by highly motivated persons.

Research on the effects of cold, summarized by Fox (1967), has concentrated on manual manipulation. Cold can alter tactile sensitivity through numbing and can degrade hand and finger dexterity; fingers become less flexible when the hand cools. A cold ambient temperature may combine with air flow (wind) to produce cooling called wind chill; though the contribution of air flow is substantial, it would not be a factor within closed industrial structures. Considerable research has examined the relationships between the ambient temperature, whole body temperature, and hand skin temperature. It is important to know which provides the best index of cold to which to relate changes in performance; hand skin temperature has been the favorite, and finger dexterity definitely deteriorates when this falls below 60°F. Tasks have included handwriting, typewriting, gear assembly, and various standardized manipulation tests; manual operations in these resemble some of the dexterity skills required in industrial plants. When other kinds of tasks have been investigated, results have been less clear-cut. Although subjects performed poorly at 40°F in a complex tracking activity, they gave evidence of inattention, impatience, and withdrawal tendencies as the intervening factors that impaired performance. Teichner (1958) found that although lightly clad or near-nude subjects reacted more slowly to simple visual signals during wind chill, their reaction times were not affected by ambient cold alone. Interestingly, in a shipboard study of watchkeeping in very cold weather (which presumably included wind), there were delays in reporting signals that Poulton (1970) said suggested some loss in brain efficiency.

Some effects of temperature on performance seem to have escaped investigation. Shivering, for example, could affect manual dexterity. Sweating might do the same due to loss of friction (and could also affect vision). Guidelines about temperature levels have been developed by the American Society of Heating, Refrigerating, and Air Conditioning Engineers and published in its *ASHRAE Guide and Data Book*. Emphasis, however, has been placed not on performance but on comfort. Standards for performance are difficult to specify because it takes so many forms and varies so extensively among individuals.

The last ambient condition in industrial plants to be considered in connection with job activities is illumination. McCormick (1970) has noted:

> There is an implicit assumption that the level of illumination for a work task may have some bearing on the performance and other criteria of

people on the task in question. The possible effect of illumination on work performance undoubtedly would depend in part upon the criticalness to the job of visual discriminations and upon the difficulty of those discriminations (size, contrast, movement, etc.). Although there have been many laboratory studies of visual performance, oddly enough there have been relatively few documented studies of actual work performance under various levels of illumination.

However, McCormick published the results of early surveys indicating increases in work output from 4% to 42% as a result of increasing illumination levels in metal-bearing manufacturing, steel machining, carburetor assembly, iron manufacturing, buffing shell sockets, piston-ring manufacturing, inspecting roller bearings, iron-pulley finishing, spinning, weaving worsted cloth, and spinning wool yarn. In most cases the original illumination was very low; in other words, during the first few decades of this century manufacturers seemed to have little understanding about the need for adequate lighting until this was demonstrated. Not all early attempts at on-site research were productive. A major investigation sponsored by the National Research Council in the 1920s was conducted in a number of laboratories and plants, including the Hawthorne facility of the Western Electric Co., but nothing more than a news report (which lacked detailed data and procedural information) was ever published (Parsons, 1974b). More recently, Sucov (1973) reviewed 16 industrial field studies in Europe. Productivity increased 10% (median) when illumination levels were raised from 15-foot candles (median) to 100-foot candles (median).

In place of on-site research, attempts have been made to apply laboratory data to the industrial scene. In the 1930s, some well-regarded illumination recommendations based on finger tension during reading were disputed as improperly calculated, specious, and excessive (Tinker, 1939). More recently, data on visual discriminations in the laboratory have been interpreted to derive illumination levels recommended by the Illuminating Engineering Society for assembly and machine-shop tasks. However, these extensions have been criticized as simplistic and unrealistic. An international group of lighting experts has been reconciling American and European viewpoints concerning both methodology and required levels.

Several approaches differ from merely extrapolating from laboratory data that indicate how much illumination is needed to make a simple visual discrimination. In one, emphasis is placed rather on the luminance (brightness) of what is viewed as the result of illumination and on the contrast in brightnesses between what must be discriminated and its background (difference in reflectances). Discrimination can be improved by increasing the contrast as well as by more intense lighting; and in high-contrast tasks more intense illumination after a point does little good. In another approach, illumination requirements are developed by investigating industrial tasks directly. A third (Faulkner and Murphy, 1973) emphasizes special-purpose lighting that can strengthen visual contrast of brightness, color, texture, or form and assist inspection tasks in industry in particular, making use of light sources with particular spectral compositions, transillumination, polarized light, spotlighting, edge lighting, black light, and other techniques. Still a fourth concern is the design of the reflecting surfaces surrounding viewers, including workers in industrial plants. With a constant amount of lighting, the more light reflected the greater the overall illumination, an important consideration with new requirements to conserve energy.

Feelings

Reactions of satisfaction and dissatisfaction or like and dislike about ambient conditions have been investigated among industrial workers through attitudinal surveys and extrapolations from laboratory findings. Some surveys have also asked how workers felt about the appearance of their surroundings, but such inquiries have not figured in the published literature. Presumably, at times the appearance of a factory's interior has entered into labor–management contract or grievance negotiations if it was so ugly or disheartening that workers expressed revulsion, or simply distaste, but such reactions must have been overshadowed by more traditional concerns about compensation, hours, and supervision. Possibly American middle-class interest in preserving the natural environment and keeping America beautiful will spread to constructed environments where blue-collar workers spend their working lives. Should plant interiors be drab and dull, or bright and exciting? Should their color schemes be enheartening or dispiriting?

One environmental aspect that has, indirectly, affected workers' feelings is spatial arrangement of machinery. As noted in the preceding section on job activities, such arrangement makes possible job enrichment. According to Lawler (1973), studies of job enrichment programs have shown that the rearrangement of machinery to broaden tasks creates greater satisfaction, at least among workers from small towns and rural backgrounds. Greater satisfaction in enriched jobs may come about, at least in part, because such jobs are less monotonous; they involve less boredom.

Among the ambient conditions that affect workers' feelings, noise can produce annoyance or irritability, but except for a German study of steelworkers there appears to have been little investigation of noise annoyance in industrial plants; the main concern has been injury to hearing. Annoyance has been studied primarily with regard to airports and aircraft, vehicular traffic, construction, and noise emanating *from* the factory into the community. Many variables make noise more annoying or less annoying to people: its intensity; its sound frequencies (very high and very low frequencies are more annoying than middle frequencies); its intermittency or aperiodicity (continuous or periodic noise is less annoying); the variability in intensity and frequencies (steady noise seems to be less annoying); its directionality (localized noise is more annoying); and its predictability (which depends on intermittency and variability). People are less annoyed when they themselves create the noise or regard it as appropriate or meaningful, that is, inherent in some worthwhile activity. They are more annoyed if it prevents or disturbs sleep or interferes with conversation or listening. They differ among themselves according to prior conditioning and they adapt. In noisy industrial plants, presumably there is much adaptation, and workers may also be affected by the task-related nature of the noise, especially if they are themselves taking part in the task. For a general treatment of noise annoyance, see Kryter (1970), who describes the various kinds of annoyance measurement procedures, e.g., ratings. Annoyance (or discomfort) can also result from vibration, but research does not seem to have encompassed this kind of vibration effect.

Loud noise has a number of physiological effects that could be correlated with feelings. These have been described by Plutchik (1959) and Kryter (1970). They include changes in heart rate and blood pressure, skin resistance, breathing, muscular tension, pupil dilation, gastric secretion, and chemical changes in blood and urine from glandular stimulation.

In contrast to noise, the feelings industrial workers have about music while they work have received a certain amount of attention. As indicated earlier, though music has not been shown clearly to increase productivity, workers in factories like it (McGehee and Gardner, 1949; Newman *et al.*, 1966). According to Uhrbrock (1961), instrumental music was preferred to vocal. Some investigators have claimed that music in industrial plants has a "calming" effect.

Discomfort can occur in industrial plants due to ambient temperature outside the "comfort zone" that has been calculated for various situations by the American Society of Heating, Refrigerating, and Air

Conditioning Engineers and set forth in their handbooks (McCormick, 1970). Factors influencing thermal comfort include air temperature, air moisture content, air pressure, air velocity, temperature and humidity fluctuations, temperature of surrounding surfaces, duration of exposure, and a worker's activity, sex, age, body build, physical condition, acclimatization, diet, expectancy, and clothing, with many interactions among these.

While discomfort can also originate from too intense general illumination, in industrial plants a more likely source is glare. This can be an unshielded lamp or a reflecting surface in the background that has too great a contrast in luminance with the immediate surround of what is being examined. Thus, it becomes important to consider lights and reflecting objects. To some extent it may be possible for illumination to produce pleasant feelings as well as disagreeable ones when it is used to emphasize scenes or objects that are enjoyable and colors that induce certain moods.

Unpleasant odors characterize the workplaces of a number of industries: pulp mills, meat rendering plants, petroleum refineries, chemical plants (e.g., phthalic anhydride and pharmaceuticals), fertilizer production, roofing materials (asphalt) production, soap and detergent manufacture, coke production, metal casting, resin manufacture, rubber processing, adhesive manufacture, pesticide production, paint and varnish production, and tanneries. While the public worries about the effects of their odors on the communities downwind, hardly a word has been said about effects on the workers inside. For example, a recent major conference on Odors: Evaluation, Utilization, and Control (Cain, 1974) entirely disregarded within-plant odors and their possible abatement. Though considerable adaptation to unpleasant odors must occur among the millions of workers in these industries, it would be interesting to know how much they dislike this kind of ambient condition in the work environment.

A worker's feelings of satisfaction or dissatisfaction about a work environment may arise in part from some of the resources or lack of resources he finds there. Researchers have investigated employee reactions to the provision, design, condition, and propinquity of washrooms, cafeterias, seating, and recreation facilities. Parking locations are also a presumed source of happiness or unhappiness.

Manipulation

Industrial workers often manipulate their work environments in their entirety. They may or may not go to work, striking, taking sick leave, or simply being absent. They may go to work late. They may give up one place of employment and go to another. What causes them to do so? Probably aspects of the physical setting make some contribution to their decisions, such as heat or cold, or some lack of protection that might make the setting dangerous or unhealthy. According to Lawler (1973), job enrichment—based on spatial rearrangement—has been shown in some studies to have reduced absenteeism and turnover. Kryter (1970) referenced one study that "found absenteeism from the work room dropped when noise level was reduced," but Broadbent and Little (1960) failed to get any reliable effects of noise reduction on turnover or absenteeism. According to Cohen (1968), "Data coupling industrial noise conditions with measures of accident rate, absenteeism, and employee turnover are not available."

Work environments are also manipulated in part. Workers move temporarily from one location to another within a work setting to get a change of scene, a change of temperature, or a change in sound effects, even if the move is from the workbench to the men's room or ladies' room. They may simply daydream, take off a shirt, or put on ear protectors. Their few options for changing major aspects of the setting itself consist of suggestions to management, union negotiations, and sabotage. Manipulating *parts* of their environments is precisely what their work consists of: operating machinery that moves, alters, or assembles materials or otherwise makes things change. Maintenance personnel manipulate the condition of machines. Inspectors change the nature of products. An industrial plant is a place where parts of the physical environment are being manipulated all the time, both resources and output.

Health and Safety

Industrial hygiene is a general term for the protection of workers' health and prevention of industrial accidents. The aspects of the physical environment in this broad and complex field can be examined here only through an overview. Studies can be found in human factors and ergonomics publications. The American Industrial Hygiene Association has a journal, annual meetings, and a large number of technical committees. The American National Standards Institute has helped develop standards. Recent federal legislation has supported workers' protection through the Occupational Safety and Health Administration (OSHA) within the Department of Labor, the National Institute of Occupational Safety and Health (NIOSH), which has sponsored accident prevention and occupational health research, and the Environmental Protection Agency (EPA).

Effects of the physical environment in industry on health and safety (see *Encyclopedia of Occupational Health and Safety* published by the International Labor Office in 1971) fall into six categories:

Cuts, Amputations, Burns, Broken Bones, and Other Bodily Harms. These result from accidents that occur (1) at machinery with moving parts; (2) in locomotion from place to place; (3) in handling hazardous materials and tools; and (4) during the movement of conveyor belts, trucks, cranes, movable platforms, etc. Environmental elements to prevent injuries from moving parts of machinery (such as punch presses, stamping presses, saws, grinders, and cutters) include guards, covers, enclosures, hoods, and barricades that prevent a hand or foot from being placed at the wrong time in a risky location (cf. the plastic hood on a pencil sharpener); the "no hands in dies" approach; devices that feed parts or materials to machinery mechanically in place of manual feeding; sensors that stop the machinery when some bodily part is in a danger zone; interlocks that prevent inadvertent or untimely activation of machinery; and remote handling devices that a worker operates in place of his arms and hands. To minimize falls, lacerations, and other injuries while workers are walking between tasks or to and from work, proper design calls for human-engineered dimensions of stairs, railings, ladders, etc.; eliminates any sharp projections, especially at corners; provides ample illumination; and prevents the use of aisles and passageways for temporary or permanent storage of objects that might become obstacles. For handling hazardous materials or tools, workers are given protective mini-environments: gloves, hard hats, goggles, or clothing. Protection from the various kinds of carriers can be provided by auditory warning signals and mechanical guards.

Explosions and Fires. Protection can be built into the physical environment in an industrial plant by proper ventilation, partitions, or remote location of explosive gases; by warning signals and alarms; by human engineering design of control panels (e.g., through coding, standardization, and placement) to reduce the likelihood of errors from misreadings of displays, misreachings to controls or accidental activation of a switch; and by minimizing distractions and maximizing the attention of operators. The environment should also be designed for rapid egress in case of emergencies as well as ingress of fire-fighting equipment.

Poisoning. Many industrial plants contain toxic elements that can poison workers through touch or inhalation: various solvents, dust and smoke, vinyl chloride in plastic fabrication, dimethyl formamide in plastic textile manufacturing (e.g., artificial leather), ozone from welding arcs, asbestos dust (producing silicosis), formaldehyde, phosgene (from burning natural gas), lead dust, and chromate (in cement). Poisoning is possible from ionizing radiation from radioactive materials or in nuclear power plants. Countermeasures, in addition to minimizing amounts and potential human contact, include protective gloves and breathing apparatus.

Ambient Conditions. Heat stress is a threat to physiological well-being in some industrial plants, especially where hard physical labor is combined with heat sources, as in foundries. Tolerance to heat stress has been determined in a number of laboratory studies. Heat prostration and frostbite can result from temperature extremes originating within or outside the plant when temperature control equipment is ineffective or protective shielding or clothing is lacking, e.g., in sheds without walls. Probably the most damaging ambient condition is noise, which at high intensities causes not only temporary deafness but gradual loss of hearing (Kryter, 1970) and possibly cardiovascular and other disabilities. In a major dispute concerning the maximum noise level that should be permitted in plants, industrial organizations, unions, and different agencies of the U. S. government have been taking contrary positions as to whether this should be 90, 85, or 80 dB(A). In an OSHA study it was estimated that 7 million workers in the United States suffer hearing losses of more than 25 dB(A) by the ages 55–59. Enforcement of the 90 dB(A) maximum would reduce this total by 10%; the 85 dB(A) maximum would reduce it by 22%. The cost in either case would be in the billions. There are various ways to make machinery less noisy. Alternatively, workers can be protected by noise shields and ear protectors. Vibration is also a threat.

Strains and Dislocations in Industrial Plants. Various muscular, joint, and spinal problems, including hernias, sprains, slipped discs, and low back pain, result from lifting, carrying, sitting, or maintaining awkward postures for extended periods. To prevent these, workers can be provided with mechanical aids for lifting and carrying, properly designed chairs for sitting, and operator stations that are designed on the basis of anthropometric and biomechanical knowledge as well as individual differences.

Safety Campaigns. In some plants safety posters (which constitute a kind of vertical litter) exhort workers to prevent accidents. Although those who design and post them may feel better for taking some action, there is no evidence they have any effect in changing workers' habits; on the contrary, habits are changed by other techniques.

Social Interaction

Workers in industrial plants interact by talking to each other on the job or during rest and meal pauses. Some on-the-job talking may be purely social (and discouraged by the management). Some may be required for doing the job: exchanging information, giving or receiving instructions, or issuing warnings, face-to-face or by telephone. Other interaction consists of body language and other signs (e.g., pointing), taking another worker's place temporarily, joint operations in some task, and sequential operations. In self-paced tasks, the speed of one operator determines the input rate for the next in line, a significant social interaction in assembly tasks. Important social interactions occur between workers and foremen. The physical setting has some effect on the level and kind of interaction by (a) keeping workers separated from each other or bringing them together through the arrangement of machinery, (b) making tasks self-paced or externally paced, and (c) providing opportunity for talking (the noise level may preclude this, for example). No systematic inquiry appears to have been directed at environmentally determined social interactions among workers in industrial plants, although a significant social life does occur (Terkel, 1974). The arrangement of machinery can be as telling a factor in social interaction at work as the arrangement of furniture is in social interaction in residences and would be as justifiable a subject for environmental design research.

Motivation

Consequences can strengthen or weaken habitual activity. Consequences intrinsic to the physical design of work settings frequently have curtailed work behavior, not strengthened it. If design makes a task effortful or a procedure inconvenient, the worker will tend to avoid it, skipping the task or modifying the procedure. On an assembly line a worker may occasionally miss an operation or make one that is incomplete. A maintenance man may fail to replace a guard on a hazardous machine, or an operator may remove it because it requires additional manipulations. An inspector may become momentarily "inattentive."

Although the physical setting provides few consequences that would strengthen desired behavior (good work habits), it could do so through information feedback about what the worker has been accomplishing, such as a tally of assembled units. This is what happened in the Relay Assembly Test Room in the Hawthorne studies (Parsons, 1974b). A related technique is to let the worker see the completed unit on which he worked (and perhaps even sign his name or at least initial it), as in successful job enrichment programs (Parke and Tausky, 1975). When the work setting is arranged so a small group of workers assembles a radio set or an automobile, the finished product constitutes a significant consequence of their actions. It strengthens these. Apparently it is this consequentiality, made possible by the physical arrangement of machinery, that accounts for higher productivity in job-enriched industrial plants, rather than the greater variety of tasks in the job. The information feedback or consequation in the form of finished product also helps make the job more "interesting" to the worker. It helps relieve boredom or monotony.

Potentiators are situations that make consequences more effective in strengthening or weakening habits. Aversive ambient conditions are potentiators. In excessive warmth a person is likely to stop working for a while for the pleasant consequence of cooling off, a drink of water, or finding a cool place. Subjects in laboratory studies have learned how to terminate noise, and the public protests aircraft noise. A worker might seek the relative quiet of the men's room to escape from factory noise. If he cannot leave the job, he may be distracted by thinking about leaving. An assembly line is an aversive situation from which there is no escape, except through daydreaming, protest, or union action about its speed. Potentiation can also consist of deprivation. If a worker is gregarious but is deprived of social interaction on the job because of the way it is arranged, he will probably take advantage of any opportunity that arises to talk to others.

Locomotion

An industrial worker may have to do considerable walking (a) to get to and from his particular workplace after reaching the plant or before leaving it, (b) from the workplace to lavatory or lunch room, or (c) in some cases between tasks that are parts of his job. Although one of the principles of "scientific management" has been to minimize a person's locomotion while he is actually at work, less concern has been indicated in industrial engineering about other locomotion, though for workers who have to walk an hour or more a day it is not a trivial matter (e.g., Terkel, 1974). Walking can be hazardous and uncomfortable as well as effortful. The extent of locomotion depends on the spatial arrangement of the plant (both horizontal and vertical) and to some degree the availability of communication circuits that make face-to-face communication unnecessary. Direct communication is required, however, for group meetings and discussion of blueprints and other documents (except where closed-circuit television, speaker-phones, and picture-phones can serve instead). Locomotion becomes critical in emergencies when workers have to get out in a hurry.

Learning

Workers get to know the place where they work. They learn how to get around in it, according to its spatial arrangement. Its appearance becomes familiar to them. They adapt to its ambient conditions. But most of their learning is related to a small part of their environment, their work station: the machinery that they operate or maintain, at which they inspect products, or between which they move parts or materials. They undergo on-the-job training whether or not this is intended by the management. They learn this mini-environment by doing, by observing others, and by getting instructions from peers or supervisors. They learn how to make fewer errors, work faster, or acquire new skills in making machinery change materials, in assembling products, or in moving elements, parts, or assemblies around. All of this is manipulation of parts of the physical setting, as noted earlier.

Perception

It is not certain from the research literature to what extent, if at all, workers comment on their physical work environments and its components, but they must at times remark on what the interior of a plant looks like and on its ambient conditions: "It looks nice and clean," or "What a racket," or "How quiet today." In factories or elsewhere a constructed environment's effects on perceptions of its users have attracted less attention than the perceptions of the designer, for example, the architect; the latter are published, with emphasis on exteriors. Nor have psychologists shown concern. They investigate perception in the abstract, and other kinds of structures (especially residences) in the particular. They obtain psychophysical judgments of sound, light, and temperature, for example, and subjective ratings of homes. But not of industrial plants.

Business Offices

An office can mean the working space (large or small) for a single individual, the collective space (large or small) for an organizational subdivision, e.g., a department, or the entire space for an organization's office workers within a building. Systematic published studies of the design of offices in any of these categories have been relatively few and have concentrated on the "landscaped" or open-plan office, although there have been some investigations of single-person offices (e.g., Fucigna, 1967), office illumination, and office equipment, such as desks and chairs.

The open-plan office is an innovation that emphasizes placing plants, bookcases, filing cabinets, or "temporary" free-standing dividers (straight or curved, and of limited height) between work spaces instead of permanent floor-to-ceiling walls. This "flexible" design has been extolled by proponents who have pointed out its lower costs but whose propaganda has otherwise outdistanced evidence of efficacy. A number of studies (Boyce, 1974; Brookes and Kaplan, 1972; Moleski, 1974; Nemecek and Grandjean, 1973; Zeitlin, 1969) have demonstrated the problems of making an adequate behavioral analysis of the change from a conventional to an open-plan office. Organizational and personnel changes may occur at the same time, confounding the effect of the change in physical surroundings. Landscaping may introduce new furniture, different or brighter colors, new lighting, and carpeting, though these could also be installed in a conventional office arrangement (except that carpeting and other sound absorption features are needed for the landscaped office). Landscaping's novelty may for a time affect users' reactions, favorably or unfavorably.

Job Activities

Activities in an office, which vary according to the job, include arranging and receiving telephone calls, talking on the phone, writing in longhand, typing, proofing, giving or receiving dictation, reproducing, filing (both input and retrieval), making computations, drafting, talking (e.g., conversing, attending discussions and conferences, conducting these, and interviewing), reading (correspondence, books, files, reports, etc.), writing, thinking (solving problems, making decisions, creating), transcribing, billing, mailing, operating business machines, operating computer terminals, and keypunching, not to mention chatting, waiting, drinking coffee, and visiting the water cooler. Task analyses of all the work done in an office could be based on observation, interviews, and documentation. Although Moleski (1974) and Zeitlin (1969) have reported studies somewhat along this line, and Fucigna (1967) another for an individual office, a complete task analysis of office work, including durations and frequencies of all activities, has not yet appeared.

How might an office environment affect the activities in it? The design could influence either the time distribution of activities or their effectiveness. The former is easier to measure than the latter. In either case measurement should be as objective as possible, but individual output can sometimes be difficult to define or record and it is also difficult to combine individual effectiveness measures into total unit measures, e.g., for a department. Alternatively one can seek certain collective measures of organizational or departmental work flow or output, hoping these properly represent the organization's entire functioning.

An alternative to getting objective data is to ask managers or staff about changes in output. Although such subjective measures are suspect, some investigators have felt they were better than nothing. Brookes and Kaplan (1972) obtained a mean subjective estimate that the landscaped office to which workers moved was less efficient than the conventional office they left. Nemecek and Grandjean (1973), on the other hand, obtained estimates indicating a rise in efficiency. In each case the respondents considered each office, new and old, in its entirety; that is, they based their estimates not on one or another aspect of the setting but on the setting as a whole. This kind of comparison left unsettled what in particular about the physical design of the office reduced or heightened the organization's productivity (if it really did change it). Further, the conventional-design offices that the open-plan offices superseded may have been inefficient or unattractive for reasons other than their physical design.

Resources. As in other work environments, office resources may or may not be provided; a resource is located in some relationship to its user; and a resource's design variously matches its user's requirements. Job activities are affected for any of these reasons. An example of an office resource is the photocopier, which reduced the clerical activity of making carbon copies or mimeographing. A copier breakdown suspends a considerable amount of office activity. Copiers have fostered such new activities as making copies of certain papers (perhaps surreptitiously) for public exposure, blackmail, or entertainment. Where the copier should be located has vexed office managers who wish to maximize both accessibility and accountability. How should copiers be designed? Should they require a standing operator? How easy is it to reload paper, insert originals, follow operating instructions, remove copies, read control settings, set controls, and clear jams? Is there ample work surface? Is the platen at a comfortable height? Are the exit trays conveniently located?

Other resources include tables, desks, chairs, sofas, filing drawers and cabinets, typewriters, telephones, storage spaces, and work and display surfaces such as drawing boards and bulletin boards. How adequately are these provided, located, and designed for office users (who include visitors and cleaning personnel as well as office workers)? The need to provide a resource becomes obvious when the telephone system is saturated or a visitor cannot sit down. Locational requirements are often unsatisfied, as evidenced by clutter, piles of paper, and makeshift displays of schedules and other information. Inadequate personal storage space has been a major problem in the landscaped office, only partially resolved by hanging shelves and placing displays on partitions. Although Brookes and Kaplan (1972) found that "adjacencies and layout" for the individual worker were a major source of dissatisfaction in both the conventional and open-plan office, it has not been demonstrated empirically how much such deficiencies affect job activities. For example, when the daily activities of three users were recorded in an old individual office and then in a new one that was supposed to have optimal locational arrangements (the "action office"), Fucigna (1967) reported "there was no measurable change in the individuals' performance, work habits or behavior which could be directly attributable to the Action Office furniture." However, the subjects expressed opinions that the new office was more efficient, minimized waste of time, accelerated information handling, reduced the forgetting of critical action items, and decreased the physical strain of doing work.

Design studies have concentrated on typewriter keyboards and chairs. Although anthropometric and biomechanical studies of chairs in offices and elsewhere have been concerned more with their comfort than effects on performance, it has been shown that optimal chair height (and desk height) can make such psychomotor activities as typing more accurate and less energy-demanding (see Burandt and Grandjean, 1963; Floyd and Roberts, 1958; Kroemer, 1971); adjustable-height chairs have accordingly been widely introduced to cope with individual differences. (Makers of piano stools knew about such things decades earlier.) Designers of open-plan offices have suggested that landscaping calls for differently designed furniture (e.g., "modular"), but it is not clear just what the differences have to be, how much they affect job activities, and to what extent they are simply changes in terminology (a desk becomes a "task response module").

Spatial Arrangement. Any office design determines the allocation of floor space to individuals, the work space, elbow room, and storage space available for each employee (i.e., the locational aspects of resources). The arrangement of the individual areas affects grouping, access, communication, interperson propinquity, noise, ventilation, and other factors that in turn influence work activities.

Individual work space is not allocated solely on the basis of what is needed for certain activities. Those higher in the hierarchy get the bigger spaces. In open-plan offices, on the other hand, those higher in the hierarchy may get the higher panels.

Though irregularity replaces rectangularity in layout, an open-plan design does not necessarily mean an entirely different spatial arrangement. The number of square feet per worker can increase, decrease, or stay the same. Proponents of landscaping usually claim a saving in space, and crowding becomes physically more feasible. Brookes and Kaplan (1972) found that crowding of desks was one of the principal objections to the landscaped office. Actually, more space is needed between individual locations to attenuate noise. Individual locations may bear the same directional relationships to each other. Personnel exercising the same function (i.e., in the same group or department) can still be situated in the same area. Small-group members can be co-located. Secretaries can be placed next to executives (a spatial arrangement that exerts considerable control over office activities), and top executives can occupy corner offices.

Although conventional offices are sometimes described as sets of cubicles of various sizes with floor-to-ceiling walls along corridors, they include the "bullpen," with scores or even hundreds of desks next to each other, row on row. This pattern, which once characterized engineering departments in the aerospace industry, typifies the work environments of large clerical staffs in banks, insurance companies, and other business concerns. For these personnel, a shift to an open-plan office means more partitioning, not less. In conventional office spaces in which many people work, bookcases and filing cabinets have functioned as dividers long before landscaping came on the scene. Conventional offices are perhaps best characterized as mixes of compartmented rectangular arrays for some workers and multidesk shared spaces for others.

Any arrangement that lacks floor-to-ceiling partitioning will permit visual and auditory distractions that interrupt job activities, especially next to circulation routes. Brookes and Kaplan (1972) said a few workers objected to traffic flow in both conventional and open-plan offices. If lack of partitioning in an open-plan office or a bullpen accounts for more errors in typing among secretaries or poorer decisions by executives, this effect has not been demonstrated, and it should be realized that distractions are momentary and that workers can adapt to them.

Ambient Conditions. Those that affect job activities in an office are primarily temperature, illumination, and noise. Some of the earlier discussion of these in industrial plants is pertinent to offices and will not be repeated here.

According to Wyon (1974), relatively recent research has shown that "moderate heat stress lowers arousal and decreases the will to work," and decrements in performance occur in tasks similar to job activities in an office. Wyon reanalyzed data from a 1923 report of the New York State Commission on Ventilation describing experiments conducted in 1913. Subjects typed and did multiplication in their heads. When they were free to type or not, they spent more time typing when the ambient temperature was 24°C than when it was 20°. Viteles and Smith (1941) reported that young men's output in mental multiplication, number-checking, and typewriter code tests deteriorated when the temperature reached 31°C (87°F). Other studies also reviewed by Duke, Findikyan, Anderson, and Sells (1967) do not seem to represent job activities in offices.

Illumination requirements in offices have been described by Hopkinson and Kay (1969). Standards have been developed by the American National Standards Institute and the Illuminating Engineering Society. European standards seem to be higher than those in the United States. Since the amount of light required depends largely on the contrast in the material viewed, office lighting must consider carbon copies, mimeographed materials, handwriting, and other matter that have relatively poor contrast. In an open-plan office or bullpen most of the lighting can be provided by ceiling-mounted luminaires. Although this saves money in fixtures, such overhead lighting can produce "veiling reflections" at documents being examined below the source, making it difficult to read ball-point or typed material. Some side-mounted lighting is desirable but difficult to create where there are few partitions.

Noise is the ambient condition having the most effect on office activities. Office noise, coming from telephone ringing, talking, and business machines, is relatively continuous or intermittent. Exterior noise—that of traffic, construction, or aircraft—can enter through open windows. In compartmented offices, noise from a particular inside source is absorbed and reflected by floor-to-ceiling walls so it cannot proceed far from the source unless the door is open. In open offices, whether landscaped or bullpen, it is attenuated only to a limited extent by screens or partitions that do not reach the ceiling (Lewis and O'Sullivan, 1974), so the attenuation occurs mostly as the sound proceeds farther from the source; "a screen is at its most effective when it casts a large acoustic shadow; that is, when it is high, wide, and the speaker is close to it." Because of the distance factor in attenuation, the farther people are apart the less noise one will receive from another. Absorptive materials are helpful, on walls, partitions, ceilings, and floors. However, ceiling absorption must be reconciled with designs to accommodate overhead luminaires. Since people are also sound-absorbent, high density would be beneficial except that with high density each individual tends to increase the amount of noise transmitted to neighboring workers.

Noise has been one of the largest problems posed by office landscaping (Brookes and Kaplan, 1972; Boyce, 1974; Nemecek and Grandjean, 1973). In addition ot its annoyance, it affects job activities through interference (masking) and through distraction. Masking means that one office worker cannot hear another because of the noise, unless the other speaks louder, which adds to the noise for other workers. Standard techniques measure the masking effects of noise through the "articulation index." Noise distraction comes primarily from intermittent or unpredictable noise, which Glass and Singer (1972) have shown affects performance in proofreading more than predictable noise. No standard technique exists for measuring distraction; "mental concentration" is difficult to describe in objective terms. Although noise can affect job performance in some other fashion when the intensity level is fairly high, one must ask how often such levels are found in offices and whether office tasks are sufficiently similar to tasks involving divided attention to suffer similarly from noise. Apparently no studies of noise effects on performance have been conducted within actual offices, although in the Weinstein (1974) investigation noted earlier one of the noise sources was teletype terminals.

Feelings

Resources. As we have seen, a major resource in an office environment is the chair. (This is becoming increasingly the case also for industrial plants; their populations are still less sedentary than those in offices, but nowadays more industrial workers sit than stand.) As a mini-environment, the chair surrounds a substantial amount of a person's anatomy. Among its physical characteristics are the depth (length), width, slope, shape, and height above the floor of the seat; the angle, height (both lower and upper parts), and curvature or fit of the back support; and the clearance for feet and calves under the chair. The principal feeling a chair can produce in a sitter is discomfort. ("Comfort" is simply the absence of discomfort.) How long one sits in a chair and what one does while sitting share in determining how much discomfort a chair arouses. A chair should be designed to facilitate changes in position and posture (Branton, 1969), although primarily chairs should make body weight impinge on the chair surface through the ischial tuberosities, with which we are endowed for this purpose. It is difficult to design a chair to suit everyone because of the anthropometric differences between individuals. Even for short-legged people a sitter's feet should be supported by the floor lest the edge of the chair cut into the thighs, a major source of discomfort. As indicated earlier, office chairs should be adjustable in height, for comfort as well as effective performance. Some chairs are designed specifically for office workers. Others are general-purpose chairs that can be used at desks. Particular design aspects can be built into chairs for visitors. For example, since more effort is required to rise from an easychair than a hard one, presumably the visitor in an easychair will be less inclined to leave the place he or she is visiting. Sommer (1974) wrote that an uncomfortable chair has been designed to keep customers of cafés from lingering in them too long. Various standards have been developed for office chairs, e.g., by the British Standards Institution.

The way to find out how uncomfortable (or comfortable) a chair seems to a worker is to ask him or her, often by administering a rating scale. Shackel, Chidsey, and Shipley (1969) installed 10 different chairs in an office and 20 users rated them for general comfort and ranked them for "body area comfort." These workers also filled out a "chair feature checklist" to indicate their reactions to particular aspects of chair design. Grandjean, Hünting, Wotzka, and Schärer (1973) asked 50 subjects to rate and rank 12 multipurpose chairs for discomfort reactions in the neck, shoulder, back, loin, buttocks, upper leg, lower leg, arms, and in general.

Ambient Conditions. Feelings of discomfort result from excess warmth or cold, poor air circulation, or drafts. Complaints about warmth and dryness, drafts, and air conditioning have accompanied some installations of open-plan offices, where it is difficult to create uniform conditions or those that please everyone. Apparently air conditioning (like overhead lighting) works better with low partitions, but high partitions are preferable for acoustical privacy. A minimum air velocity of 10 feet per minute has been recommended to avoid stagnant zones. Although "freshness" of the surrounding air matters to many people, it has received little systematic scrutiny. Bedford (1961) associated air speeds with "arousal," which was related also to relative humidity (a dry atmosphere being viewed as more stimulating) and mean radiation temperature. It has been pointed out (Hanes, 1974) that comfort or discomfort sensations from the surrounding temperature depend not only on the temperature and humidity but also on the activities of the individual, his or her clothing, and his or her diet. Typing produces only a little more heat from the individual than sitting still, about the same as standing. In offices, clothing is likely to be relatively light, except when the temperature drops and heating fails. Diet may be reduced for some individuals. Since there are extensive differences between individual comfort reactions to temperature, recommendations take the form of a range or "comfort zone," mentioned earlier, which still cannot suit all office employees. Everyone seems to have some capability of accommodating to some fluctuations in temperature through partial adaptation as well as by wearing different clothes (Humphreys and Nicol, 1970).

Office illumination can also affect workers' feelings. As in factories, glare that produces discomfort comes either from an unshielded lamp or from too great a contrast between the brightness of the immediate work surface, such as paper being typed, and the surroundings; these should not be more than three times brighter or darker. Feelings of pleasantness, or perhaps unpleasantness, are often associated with the nature of the illumination. Many persons prefer daylight to artificial light and thus desire offices with windows. Daylight enters the room horizontally and thereby gives objects a somewhat different appearance from that produced by overhead lighting. It may also have some connotations of linkage with the outside world. Even when offices do provide daylight, artificial illumination is usually combined with it, more being needed the farther the worker is from the window. Individuals distant from a window overestimate the extent of the daylight component, perhaps because they like it. According to a study by Harper (1974), people vary greatly in their feelings toward different kinds of fluorescents. Helson and Lansford (1970) found that individuals enjoyed a distinct brightness contrast between object colors and background colors (though this should not be so great as to produce the glare effect noted above). Filters substantially altering spectral composition of illumination can, of course, affect surface hues, on walls or working surfaces, but reactions to the actual surface colors are little influenced by the type of illumination, e.g., incandescent versus fluorescent. It has been suggested that people prefer variety in lighting, though it may be difficult to satisfy this preference in open-plan offices due to the relatively uniform lighting from luminaires. Brookes and Kaplan (1972) found numerous dislikes for the illumination in both open-plan and conventional offices but more likes than dislikes in each.

Like industrial workers, office workers are often annoyed by noise, but the nature of the noise in offices is different from that in industrial plants because the primary sources are office machines and other human beings talking. The office employees surveyed by Brookes and Kaplan (1972) listed "noise of conversations" as the most disliked aspect of their open-plan office and as the second most disliked aspect of their conventional office. Noise from office machines and telephones produced less annoyance, probably because this noise was viewed as job-related, i.e., "necessary" noise. Systematic investigations of noise annoyance in offices have been few despite its importance, and there has been no effort to distinguish the feelings that result directly from the noise from those that occur because the noise has interfered with speech; instead, noise has been characterized simply as "bothersome" or "disturbing." More data are needed about the characteristics of open-plan office noise such as frequency spectrum, continuity, intensity, and unexpectedness, to evaluate the effectiveness of such proposed countermeasures as masking by white noise.

Appearance. An office's appearance depends on its decor, furnishing, arrangement, lighting (already discussed), and other factors, which contribute to the appearance's effects on users' feelings. Since most open plan offices emphasize all these aspects, it is not surprising that office workers have given office landscaping high ratings and complimentary remarks for their "colorful" and "cheerful" appearance or atmosphere and "attractive decor" (Brookes and Kaplan, 1972). With their low and interrupted barriers it is possible to get a visual overview in open-plan offices that is denied in compartmented offices, and the irregular arrangements and methods of demarcating spaces differ from the bullpen. Zeitlin (1969) also found feelings of user satisfaction about the appearance of landscaping—no doubt a major factor in its popularity, which has been only slightly lessened by aversions to the insects that sometimes come with the greenery. A shift to a landscaped office makes it possible to install new and perhaps more appealing furniture of all kinds and introduce lighter colors, which may affect the moods of both the decision maker who arranges the shift and the workers involved.

Although individuals differ in their color preferences, and tastes change over time, contemporary preferences have been surveyed by Helson and Lansford (1970) and Guilford and Smith (1959). According to the latter, for most values of saturation and brightness, both of which influence how much a person likes a hue, greens and blues are preferred to yellows or yellowish green, a finding with which the former authors agreed; these authors concluded that women prefer "warmer" colors and men "cooler" colors. However, it is not certain how extensively preferences ascertained through samples can be extrapolated to walls, floors, and ceilings, where there are likely to be combinations of colors. Various contrasts affect feelings in different ways, as pointed out by these and other authors (e.g., see Mehrabian and Russell, 1974).

How important are the feelings of office users about the appearance of their work environment? Sommer (1974) asserted that "people should have the right to attractive and humane working conditions," despite managements that want to color all walls gray or institutional green (a federal government favorite whose introduction was regarded as a vast improvement over the somber hues of earlier days). Although economics, efficiency, or just plain insensitivity may favor standardization, Sommer issued a call for variety. Despite the cost, some office workers, chiefly executives, are able to achieve the aesthetic satisfactions they desire. What about the rest? Should serious attempts be made to learn their likes and spend the money to satisfy these, whether or not their enjoyment of their surroundings affects their working effectiveness?

Manipulation

Office workers, like industrial workers, react to their work environments in their entirety by going or not going to work, arriving and leaving early or late, and changing places of employment. How much of this manipulation results from the physical environment and could be changed by altering that environment cannot be estimated, but as in the case of industrial plants, the setting undoubtedly makes some contribution of these kinds of behavior.

Users of offices can occasionally manipulate them by choosing or altering some of their resources. Though this is especially the case with those in higher echelons who are able to select or change their own furniture, Sommer (1974) has suggested:

> In a large corporation or agency it would be feasible to give each employee
> a choice of desk, chair, file cabinet, table, and waste basket from a central
> furniture pool, not only at the time of employment but every six months if
> the person felt like changing things around.

Many workers, especially those in small offices, manipulate resources by arranging their furniture—desk, table, file, bookcase etc.—as they choose, though clerical workers have less latitude in doing so. Since in open-plan offices bookcases and files often function as dividers, this "flexible" design may actually permit less variation in the arrangement of individual work stations. On the other hand, special provision may be made to hang files and shelving on low-height partitions according to individual requirements, a manipulation that some managements forbid in the case of floor-to-ceiling walls.

It is the overall spatial arrangement that has earned office landscaping its reputation for flexibilty, because the elements that separate small work areas can be readily moved from place to place at low cost. Undoubtedly such flexibility has been exploited when organizations have gone through reorganization or staff size has changed. Thus, the management has been able to manipulate the physical enviroment in an open-plan office more easily than in a conventional office. At the same time one might ask whether this has occurred as often as it should, since even temporary structures acquire a kind of permanence. How much overall rearrangement has occurred because office personnel have themselves wanted it?

Office workers alter the appearance of the places where they work in other ways. Though they may be unable to repaint the walls, they put up posters and pictures, bring in plants and flowers, and even post interesting items on corridor walls. Thus, they try to personalize their environments to the extent they can.

They also manipulate their settings by altering ambient conditions, turning lights on and off, opening or closing windows or drapes, changing air conditioning and temperature controls, and closing or opening doors to vary the noise level. The extent of this kind of behavioral reaction to the environment depends, to be sure, on the availability at the work location of a light switch, window, temperature control, or door. When they can, office workers also manipulate their environments to facilitate particular social situations or interactions, for example, to achieve privacy or confidentiality. Overall office design, such as landscaping, can significantly affect individual options for all these manipulations.

Health and Safety

The ambient conditions that threaten health in industrial plants are generally absent in business offices. Temperature seldom reaches a level to bring heat prostration or noise an intensity to damage hearing. Chemical contaminants do not create toxicological hazards. Nor are there major dangers of accidents, other than fire, which in office buildings can trap workers on upper floors. To some extent office employees need protection from their fellowmen more than do industrial workers. Burglars and robbers find easy prey in offices, especially at night (as at the Watergate) or on paydays. As a result, the premises may have to be protected through such physical safeguards as special doors, barriers, and reception areas. Within offices, lack of barriers (as in bullpens or landscaped offices) may encourage thefts by fellow workers and anxieties about such thefts.

Social Interaction

Social interaction characterizes organizational functioning in business offices, involving dyads, triads, or small groups. It consists of visual contact, talk (direct or by telephone), and to some extent touch (including indirect touch, as in handing someone a piece of paper). It includes interaction between workers at the same (more or less) organizational level, those at different levels (e.g., superior and subordinate), and employees and visitors. Most office interactions are work-oriented, such as exchanging information (e.g., asking or answering questions) and supervision (e.g., giving instructions or getting reports). Direct talk often requires movement (locomotion) of one or more persons to come closer. Many office interactions are not work-oriented but rather of an informal or social nature. All of the foregoing is obvious to those who work in offices. Perhaps it is because they are so obvious that office interactions have neither been codified in task analyses nor closely investigated by social or organizational psychologists. Yet, they should be specified to determine how a particular physical environment affects social interactions through its spatial arrangement, communication facilities, and resources.

Spatial Arrangement. According to the overall spatial arrangement individual work spaces are closer to each other or further apart, in gradients of proximity. On the other hand, walls and other barriers separate them. Barriers inhibit some interactions by separating individuals from each other (visually, acoustically, and locomotionally) and they increase other interactions when they enclose dyads, triads, or a small group and inhibit contact with others outside the enclosure. Among the outcomes of proximities, barriers, and their interrelationships in business offices are variations in privacy, personal space, and territoriality (see Altman, 1975, for analyses of these terms), as well as variations in supervisory effectiveness, organizational awareness, cohesiveness or group identity, sociability, and status.

In business offices, privacy—visual or acoustical—is desired by many workers, especially those in middle or upper echelons, who according to surveys have tended to show more resistance to office landscaping than lower-echelon workers. (In the move from the conventional office, the middle and upper managers move out of private or semiprivate compartments into more public spaces; the clerical staff moves from one open space, e.g., a bullpen, to another that may have more dividers.) In an open-plan office, even with head-height partitions, everyone can see whether or not someone is at work, present or absent, writing a report or reading the newspaper, thinking or catching forty winks, typing or with feet on desk. Everyone can see who is talking (directly) with whom. Conversational confidentiality, especially in person-to-person talk but also to some extent in phone conversation, suffers when floor-to-ceiling walls give way to lower partitions or no barriers at all except plants and files. The loss of privacy in general and acoustical privacy in particular has been a leading complaint about office landscaping (e.g., Brookes and Kaplan, 1972). As pointed out in connection with the masking effects of office noise on job activities, low partitions do not have much attenuating effect on the transmission of conversations in open-plan offices, and the primary agent of attenuation is distance from the source (Lewis and O'Sullivan, 1974). Voices carry, and without walls to attenuate sound, people have to raise their voices to be heard above the ambient noise level. Whether or not they can then be heard and understood at a farther distance than can conversations at normal or constant loudness in a quieter ambience, speakers often feel they are transmitting better to unintended recipients when they raise their voices.

Along with rises in hemlines, modesty panels have been extensively introduced for female workers in open offices. This physical alteration of the setting to cope with a particular kind of social interaction might variously be associated with territoriality, personal space, and visual privacy.

Altman (1975) has explicitly asked whether office designers consider "personal space" in planning their layouts. How close can two desks be placed to each other without each user feeling an intrusion on her or his social-distance zone? Since personal space involves angle of orientation as well as proximity, should adjoining desks be placed side by side, facing each other, at 90° angles, or how? A cellular office intrinsically denotes the "territory" of the occupant, implying ownership and control. In a landscaped office, how effective are "markers" such as plants or dividers in implementing territorial behavior? For example, will a worker greet a visitor or fellow employee with "Come into my office" if he or she has no door to open and only some shrubbery to indicate it is "his" or "hers"?

Visual privacy—or access—is a two-way operation. If a supervisor's station is situated in the same open area as the desks of his or her clerical staff, he or she can be observed but at the same time can observe the workers. It has long been assumed that continuous visual access to clerical staffs helps supervisors make sure the staffs keep working at their assigned tasks. Presumably, each worker refrains from idle conversation, daydreaming, visits to the lavatory, or otherwise goofing off because he or she knows the supervisor is watching.

Proponents of the landscaped office have alleged it would increase organizational awareness and understanding among office workers, presumably because everyone could see who everyone was and what they were doing, across vertical and horizontal boundaries within the organization. Whether this is indeed the case is difficult to discover, if only because it is first necessary to develop adequate operational definitions of organizational awareness and understanding, as well as ways of measuring them. It is similarly difficult to determine whether a small group of employees acts more cohesively or less cohesively when it works together in a room that has floor-to-ceiling walls (or in adjoining small offices) than in a shared area in a landscaped office, though some efforts have been made to measure group identity or identification. Such terms must be operationalized, as must "interaction" and "communication." Zeitlin (1969) and Nemecek and Grandjean (1973), respectively, were told by office workers that these increased when they moved to the landscaped office.

The office workers investigated by Brookes and Kaplan (1972) rated the landscaped office more "sociable" than their earlier conventional office, presumably because of its spatial arrangement. In both bullpen and compartmented offices, social interactions occur extensively in the corridors. It has been suggested that these should be designed with a width that permits people to stop and chat while at the same time others can go past. Some organizations have installed employees' lounges for the same purpose. Locations of drinking fountains and coffee bars are thought to influence the amount of social contact among office workers.

Spatial arrangement of offices has always been associated with the social phenomenon of status. The location and size of an office have been status symbols in an organizational hierarchy (Peter, 1972), as epitomized in the corner office and the executive suite. As noted earlier, in the landscaped office the height of partitions can function as a further status symbol. Height within a building can be the analog of eminence in the hierarchy, from the basement to the penthouse.

Communication. Office design incorporates both features for mediated communication and those that facilitate face-to-face communication. Mediated communication takes place mostly by means of pieces of paper and telephones, which include picture-phones and speaker-phones. An individual office location has one or multiple extensions, and a single phone has one or multiple incoming and outgoing lines. A most significant feature of a large office is the telephone switchboard and its design. Phones may be situated in different locations with attachments to provide voice privacy. All of these features affect, in one way or another, telephonic behavior. Communication by paper (memoranda, bills, correspondence, reports, etc.) involves less elaborate physical paraphernalia: messengers' carts, pneumatic tubes, mail chutes, and in-baskets and out-baskets. All these devices figure in an office's incessant information flow, which is largely social interaction.

Face-to-face voice communication has already been discussed in connection with noise interference (masking) in job activities and the effects of spatial arrangement on acoustical privacy. The loss of privacy that inhibits communication can make itself manifest in various ways. For example, if one employee hears another in an adjoining office, he or she may presume his or her own voice is being heard by his or her neighbors.

For both employees and visitors, finding one's way to someone's office or location can be difficult unless one is aided by posted diagrams or maps, office labels and numbers, and coding to distinguish different parts of a large office or office building. This kind of communication is frequently inadequate. Visitors often have as difficult a time finding their way out as locating the person they want to see. Some designers apparently feel that suitably posted information for guidance would spoil the decor. Only the fire department can override such constraints; people must know how to evacuate a building. Many visitors have to be guided in and out by the secretary of the person they are visiting, an additional social interaction some seem to enjoy. Names on office doors are often more likely to communicate the rank of the occupant than simply identify him, the method ranging from a name painted on a glass panel or printed on a card, to the name in gold leaf or engraved on a brass plate (Peter, 1972).

Resources. The furniture at an individual work location can affect social interactions according to the relative positions of individual elements. Sommer (1974) described the various patterns he found in faculty office rooms:

> One man has placed his bookcase between his desk and the door for maximum privacy. Another has joined together his desk and table to yield a large work area and writing surface as well as considerable distance from any visitor, and a third has placed all furniture against the walls to remove any barriers between himself and the students.

Altman (1975) has also remarked on such patterns:

> A formal arrangement with maximum distance and strong barriers involves across-the-desk seating of visitors. A less formal arrangement would be to have visitors seated at the side of the desk, and an even more informal arrangement would be to have visitors sit on the same side of the desk, at a neutral table, or around a coffee table. Thus it is possible to have a variety of seating configurations that can be suited to a particular relationship and to the expected nature of conversations. My own office has two such configurations—across the desk and around a coffee table.

Although seating configurations in offices have not received the same study they have elsewhere, as in restaurants and libraries, both office users and the two social psychologists just cited have presumed they would have certain effects on visitors.

Motivation

Much of what was said about motivation in industrial plants is applicable to business offices and need not be repeated here. If resources require effort or if ambient conditions are unpleasant (the potentiation), workers will tend to stay away from or leave the office (avoidance or escape being the consequence). As mentioned earlier, reactions to chair discomfort exemplify such motivated behavior. If the entire setting is disagreeable, similar behavior may ensue, although it may be called alienation and the social circumstances of the setting may greatly outweigh the physical circumstances.

It is possible to conceive of methods whereby the physical attributes of offices could be made to function as positive consequences that would strengthen certain behavior, such as coming to the office early (or at all) or lingering there after the official closing time. It would be interesting to see whether the attractive appearance of landscaping has led to any reduction in absenteeism, turnover, or unpunctuality. Effective work might be rewarded by giving the individual a more desirable location or set of furniture. Options for personalization of resources might be made contingent on work performance. Access to an employees' lounge might be arranged to reinforce productivity. In a sense, management already consequates effective performance by moving outstanding performers to bigger offices with greater prestige along with increases in pay and power. Because of these other, even more influential consequences, it is difficult to estimate how much incentive is provided simply by the new office.

Job enrichment can be instituted in business offices as well as in industrial plants, especially where clerical tasks are also assembly tasks. For example, each clerk compiling directories for Indiana Bell Telephone was shifted from handling small parts of the task to compiling an entire section of the directory, and errors dropped drastically. Although in this instance it is not clear whether the physical setting was modified to permit this job enrichment, in another case (Ford, 1973) the desks of telephone company typists were shifted to a wagon-wheel arrangement that helped each typist handle installation and repair orders for a particular geographical area; service representatives sat at the center. More service orders were processed (and turnover among typists dropped) in comparison with the number when the typists were simply in a typing pool. As in the directory csase, each typist received more feedback about her work.

In open-plan offices and bullpens, not just the supervisors and supervised but also those doing the same work can observe each other more easily. Such observation is a prerequisite to modeling, also called observational learning, where one individual imitates another because of the consequences the other receives. Consequation is vicarious, strengthening or weakening similar behavior. As a result of this motivational situation, individuals might perform more effectively because they saw others being rewarded for greater productivity; or, they might *fail* to work harder because no one else did so or no other workers were praised or paid for doing so. In addition, eye contact (as attention) can function directly as a motivating consequence, whether the contact occurs with a supervisor or an associate. These are all matters that deserve empirical investigation in office settings.

Locomotion

Those who plan business offices give considerable attention to locomotion. There must be suitable corridor or hallway width and access to lavatories and canteens as well as between parts of the organization. As noted earlier, locomotion is important for both work-oriented and informal social interactions, and communication devices are needed to inform people how to get where they want to go.

Though locomotion has its social functions, designers understandably try to minimize it. The farther apart two office workers are, the less likely one will walk or ride in an elevator to meet with the other. Walking requires effort and waiting for an elevator is inconvenient. Since walking up stairs requires more effort than walking on a level, an office worker will be less likely to locomote by stairway to join another on the next floor than to locomote to join one on the same level, though it may take even longer to reach him. These considerations mean putting close together, on the same floor, those parts of an organization whose members should be meeting together frequently.

A shift from a compartmented office to an open office, bullpen or landscaped, might result in more walking by workers who want to talk to others, on business or otherwise, because the cues for walking are stronger. If another worker is visible, it is easy to see whether someone else is already with him or her, or whether the other worker is even at his or her desk. One can observe when a visitor leaves the other worker and then get up and walk over. It should be relatively easy to get empirical data on the extent to which the spatial arrangement of open-plan offices affects the frequency and extent of locomotion.

Learning

Office workers have to learn their physical setting and its ingredients. They usually do this rapidly, and the problems of learning become apparent only during fire drills or an actual fire, or when visitors become lost. Learning in large part is the acquisition of habits. Office workers acquire the habit of going from their office to the street level by elevator, even when they are on the second floor. When a fire breaks out, they rush to the elevator and are trapped. Fire drills are needed to create a different habit for emergencies. Visitors have to learn to find their way through a complex office, just as workers must learn to navigate through the office area of another department even though they have mastered their own space. The communication aids for locomotion mentioned earlier both substitute for learning and facilitate it.

Employees also adapt to their physical surroundings. They get accustomed to appearance and ambient conditions. Reactions to open-plan offices should be investigated after the novelty of the move has worn off as well as shortly after the change.

Perception

Perceptual behavior is greatly affected by novelty. People talk a lot about what they see in open-plan offices, especially soon after they move in, or if they are visitors. But because of the extended vistas and techniques for demarcating work areas, landscaping an office may permanently produce a greater variety of interesting things to look at, and perceptual activity may remain at a higher level than that in a conventional office, something that could be determined through frequency counts of comments about plants, objects and people in the environment. The irregularity of landscaping may encourage perceptual activity, in contrast to the rectangularity of bullpen and cellular offices. Some efforts have been made to get perceptual reactions about open-plan and conventional offices from workers who have moved to the former (e.g., Brookes and Kaplan, 1972), but these have consisted of ratings on such scales as "orderly," "open," "spacious," "colorful," "curved," and "angular." In one study (Zeitlin, 1969), workers felt the landscaped office was more beautiful than it was ugly, a dimension that comes closer to an index of feelings than of perception (as defined in this chapter).

Although no one has tried to evaluate offices, conventional or landscaped, in terms of their "information" content, Kaplan (1973) offered a number of guidelines to keep offices from being "boring": Lighting should create variety; an office should not be overly quiet; layout patterns should have some irregularity; colors should be selected for subtlety and harmony; offices should include art; executive offices should show individuality; public rooms should have intimacy and warmth; and offices should seem spacious. In a sense, "boring" is the antithesis of "arousing," and Mehrabian and Russell (1974) have suggested that environments should be investigated for the degree they contribute to "arousal."

CONCLUSION

Modern technology has produced physical environments that vary in their entirety, in their resources, in their spatial arrangement, in their ambient conditions, in their communication, in their consequation and potentiation, in their protection, and in their appearance. All of these aspects of a constructed environment affect its users: Their activities, feelings, manipulation of the environment, health and safety, social interaction, motivation, locomotion, learning, and perception. Nowhere is this more evident than in the world of work where technology dominates the setting.

The analysis of two work environments, industrial plants and business offices, has illustrated the various ways in which the various aspects of an environment affect its users. The same kind of analysis based on the same sets of categories can be applied to other work environments and, indeed, to all constructed environments. Such analysis serves several purposes:

1. It should help establish a framework within which to study the effects of constructed environments on people. Such a framework seems badly needed for those who wish to pursue such study.

2. It should bring some coherence to the debate about the extent to which a constructed environment influences what human beings do, feel, and think. Although it does not specify the extent, it suggests the different paths to be followed to find out.

3. With the accumulating knowledge that is thereby systematized concerning environmental effects on people, it will be easier to design environments to benefit their users. The analysis itself can suggest what designs might be advantageous.

Due to limitations on length, this chapter has many gaps. The coverage of industrial plants and business offices is far from complete. No attempt has been made to analyze the effects of the physical environment on workers in other settings. It has not been possible to give sufficient emphasis to the many ways and great extent by which users of any work environment differ among themselves. As a result of such differences, a particular aspect of some work environment will have a different effect on the activities or feelings or health of one worker or category of workers than it will on those of another.

In trying to specify the effects of work environments on people, we must, then, consider and combine four sets of categories:

- Aspects of all constructed environments.
- Ways in which these affect human beings.
- Varieties of physical settings.
- Varieties of human beings.

REFERENCES

Altman, I. *The environment and social behavior. Privacy, personal space, territory, crowding*. Monterey, Calif.: Brooks/Cole, 1975.

Atherley, G. R. C., Gibbons, S. L., and Powell, J. A. Moderate acoustic stimuli: The interrelation of subjective importance and certain physiological changes. *Ergonomics*, 1970, *13*, 536–545.

Ayoub, M. M. Work place design and posture. *Human Factors*, 1973, *15*, 265–268.

Barnes, R. M. *Motion and time study*. 3d ed. New York: Wiley, 1949.

Bedford, T. Researches on thermal comfort. *Ergonomics*, 1961, *4*, 289–310.

Berenson, P. J., and Robertson, W. G. Temperature. In J. F. Parker, Jr., and Vita R. West (Eds.), *Bioastronautics data book*. 2d ed. NASA SP-3006. Washington, D.C.: U.S. Government Printing Office, 1973.

Blood, M. R., and Hulin, C. L. Alienation, environmental characteristics, and worker responses. *Journal of Applied Psychology*, 1967, *51*, 284–290.

Boyce, P. R. Users' assessments of a landscaped office. *Journal of Architectural Research*, 1974, *3(3)*, 44–62.

Branton, P. Behavior, body mechanics and discomfort. *Ergonomics*, 1969, *12*, 316–327.

Broadbent, D. E., and Little, E. A. J. Effects of noise reduction in a work situation. *Occupational Psychology*, 1960, *34*, 133–140.

Brookes, M. J., and Kaplan, A. The office environment: Space planning and affective behavior. *Human Factors*, 1972, *14*, 373–391.

Burandt, M., and Grandjean, E. Sitting habits of office employees. *Ergonomics*, 1963, *6*, 217–228.

Cain, W. S. (Ed.). Odors: Evaluation, utilization, and control. *Annals of the New York Academy of Sciences*. Vol. 237. New York: New York Academy of Sciences, 1974.

Carroll, D. Physiological response to relevant and irrelevant stimuli in a simple reaction time situation. *Ergonomics*, 1973, *16*, 587–594.

Cohen, A. Noise effects on health, productivity, and well-being. *Transactions of the New York Academy of Sciences*, 1968 *30(7)*, 910–918.

Corso, J. F. The affects of noise on human behavior. Report WADC 53-81. Wright-Patterson Air Force Base, Ohio: Wright Air Development Center, 1952

Damon, A., Stoudt, H. W., and McFarland, R. A. *The human body in equipment design*. Cambridge, Mass.: Harvard University Press, 1966.

Duke, M. J., Findikyan, N., Anderson, J., and Sells, S. B. *Stress reviews: II. Thermal stress—heat*. Technical Report No. 11. Fort Worth, Texas: Institute of Behavioral Research, Texas Christian University, 1967.

Faulkner, T. W., and Murphy, T. J. Lighting for difficult visual tasks. *Human Factors*, 1973, *15*, 149–162.

Floyd, W. F., and Roberts, D. F. Anatomical and physiological principles in chair and table design. *Ergonomics*, 1958, *2*, 1–16.

Ford, R. N. Job enrichment lessons from A.T.&T. *Harvard Business Review*, 1973 (May–June), *50*, 65–69.

Fox, W. F. Human performance in the cold. *Human Factors*, 1967, *9*, 203–220.

Fucigna, J. T. The ergonomics of offices. *Ergonomics*, 1967, *10*, 589–604.

Gilbreth, F. B. *Motion study*. New York: Van Nostrand, 1911.

Gilbreth, F. B., and Gilbreth, L. M. A fourth dimension for measuring skill for obtaining the one best way. *Society of Industrial Engineering Bulletin*, 1923, *5(11)*.

Glass, D. C., and Singer, J. E. *Urban stress*. New York: Academic Press, 1972.

Grandjean, E., Hünting, W., Wotzka, G., and Schärer, R. An ergonomic investigation of multipurpose chairs. *Human Factors*, 1973, *15*, 247–255.

Grether, W. F. Vibration and human performance. *Human Factors*, 1971, *13*, 203–216.

Guilford, J. P., and Smith, P. C. A system of color preferences. *American Journal of Psychology*, 1959, *72*, 487–502.

Hanes, L. F. Thermal comfort in dwellings. Proceedings of annual meeting of the Environmental Design Research Association, 1974.

Harper, W. J. On the interpretation of preference experiments in illumination. *Journal of the Illuminating Engineering Society*, 1974, *3*, 157–159.

Helson, H., and Lansford, T. The role of spectral energy of source and background color in the pleasantness of object colors. *Applied Optics*, 1970, *9*, 1513–1562

Hopkinson, R. G., and Kay, J. D. *The lighting of buildings*. New York: Praeger, 1969.

Humphreys, M. A., and Nicol, J. F. An investigation into thermal comfort of office workers. *Journal of the Institution of Heating and Ventilating Engineers*, 1970, *38*, 181–189.

Huntington, E. *Civilization and climate*. 3d ed. New Haven, Conn.: Yale University Press, 1924.

Kaplan, A. Are your offices boring? *Business Management*, 1973.

Kroemer, E. K. H. Seating in plant and office. *American Industrial Hygiene Association Journal*, 1971, *32*, 633–652.

Kryter, K. D. *The effects of noise on man*. New York: Academic Press, 1970.

Lawler, E. E. Job design and employee motivation. *Personnel Psychology*, 1969, *22*, 426–435.

Lawler, E. E., III. *Motivation in work organizations*. Monterey, Calif.: Brooks/Cole, 1973.

Lewis, P. T., and O'Sullivan, P. E. Acoustic privacy in office design. *Journal of Architectural Research*, 1974, *3(1)*, 48–51.

McCormick, E. J. *Human factors engineering*. 3d ed. New York: McGraw-Hill, 1970.

McGehee, W., and Gardner, J. E. Music in a complex industrial job. *Personnel Psychology*, 1949, *2*, 405–417.

Mehrabian, A., and Russell, J. A. *An approach to environmental psychology*. Cambridge, Mass.: MIT Press, 1974.

Moleski, W. Behavioral analysis and environmental programming for offices. In J. Lang, C. Burnette, W. Moleski, and D. Vachon (Eds.), *Designing for human behavior: Architecture and the behavioral sciences*. Stroudsburg, Pa.: Dowden, Hutchinson and Ross, 1974.

Murrell, K. F. H. *Ergonomics. Man and his working environment*. London: Chapman and Hall, 1969.

Muther, R. Plant layout. In H. B. Maynard (Ed.), *Industrial engineering handbook*. New York: McGraw-Hill, 1956.

Nemecek, J., and Grandjean, E. Results of an ergonomic investigation of large-space offices. *Human Factors*, 1973, *15*, 111–124.

Newman, R. I., Hunt, D. L., and Rhodes, Fen. Effects of music on employee attitude and productivity in a skateboard factory. *Journal of Applied Psychology*, 1966, *50*, 493–496.

Parke, E. L., and Tausky, C. Need theory, reinforcement theory, and job enrichment. *Personnnel*, 1975.

Parsons, H. M. Human factors in the constructed environment. Paper presented at the Engineering Foundation Conference on The Constructed Environment with Man as the Measure, Monterey, Calif., 1974(a).

Parsons, H. M. What happened at Hawthorne? *Science*, 1974, *183*, 922–932(b).

Peter, L. J. *The Peter prescription*. New York: William Morrow, 1972.

Plutchik, R. The effects of high intensity intermittent sound on performance, feeling, and physiology. *Psychological Bulletin*, 1959, *56*, 133–151.

Poulton, E. C. *Environment and human efficiency*. Springfield, Ill.: Charles C. Thomas, 1970.

Roethlisberger, F. J., and Dickson, W. J. *Management and the worker*. Cambridge, Mass.: Harvard University Press, 1939.

Shackel, B., Chidsey, K. D., and Shipley, Pat. The assessment of chair comfort. *Ergonomics*, 1969, *12*, 269–306.

Smith, K. R. Intermittent loud noise and mental performance. *Science*, 1951, *114*, 132–133.

Smith, W. A. S. Effects of industrial music in a work situation requiring complex mental activity. *Psychological Reports*, 1961, *8*, 159.

Sommer, R. *Tight spaces. Hard architecture and how to humanize it*. Englewood Cliffs, N.J.: Prentice-Hall, 1974.

Sucov, E. W. European research. *Lighting Design and Application*, 1973, 39–43.

Terkel, S. *Working*. New York: Pantheon, 1974.

Theologus, G. C., Wheaton, G. R., and Fleishman, E. A. Effects of intermittent, moderate intensity noise stress on human performance. *Journal of Applied Psychology*, 1974, *59*, 539 547.

Teichner, W. H. Reaction time in the cold. *Journal of Applied Physiology*, 1958, *42*, 54–59.

Tinker, M. A. Illumination standards for effective and comfortable vision. *Journal of Consulting Psychology*, 1939, *3*, 11–20.

Turner, A. N. and Lawrence, P. R. *Industrial jobs and the worker*. Boston: Harvard University School of Business Administration, 1965.

Uhrbrock, R. S. Music on the job: Its influence on worker morale and production. *Personnel Psychology*, 1961, *14*, 9 38.

Van Cott, H. P., and Kinkade, R. G. (Eds.). *Human engineering guide to equipment design*. Rev. ed. Washington, D.C.: Government Printing Office, 1972.

Viteles, M. S., and Smith, K. R. *A psychological and physiological study of the accuracy, variability, and volume of work of young men in hot spaces with different noise levels*. ASHVE Report. Washington, D.C.: Bureau of Ships, U.S. Navy, 1941.

Warner, H. D., and Heimstra, N. W. Effects of intermittent noise on visual search tasks of varying complexity. *Perceptual and Motor Skills*, 1971, *32*, 219–226.

Weinstein, N. D. Effect of noise on intellectual performance. *Journal of Applied Psychology*, 1974, *59*, 548–554.

Wyatt, S., and Langdon, J. N. *Fatigue and boredom in repetitive work*. I.H.R.B. Report 63. London: His Majesty's Stationery Office, 1937.

Wyon, D. P. The effects of moderate heat stress on typewriting performance. *Ergonomics*, 1974, *17*, 309–318.

Zeitlin, L. R. *A comparison of employee attitudes toward the conventional and the landscaped office*. Report. New York: Port of New York Authority, 1969.

6

Behavioral Ecology, Health Status, and Health Care: Applications to the Rehabilitation Setting

EDWIN P. WILLEMS

This paper is about a view of behavior–environment relations called behavioral ecology, about health status and health care for persons hospitalized with spinal cord injuries, and about some relationships between these two areas. One major purpose is to communicate some of the essential flavor of behavioral ecology. This will be done by presenting some background comments in the next section and presenting some key aspects or guiding assumptions of behavioral ecology in a later section.

The second major purpose is to present a behavioral picture of the aftermath of spinal cord injury and rehabilitation and the ways in which behavioral ecology adds clarity to that picture. This will be done by presenting some general background on spinal cord injury and then describing some key aspects of a program of research on persons with spinal cord injuries. The order of presentation in the paper will be (a) introductory comments on behavioral ecology, (b) introductory comments on behavioral aspects of spinal cord injury, (c) description of the setting and methods of the research program, and

EDWIN P. WILLEMS · Department of Psychology, University of Houston, Houston, Texas, and Texas Institute for Rehabilitation and Research, Houston, Texas.

(d) presentation of eight key aspects of behavioral ecology and the manner in which each is manifested in or exemplified by a key aspect of the research program.

It is impossible to present either behavioral ecology or the research program in a definitive manner in one paper. Interested readers will have to consult other cited literature to get those complete pictures. Nonetheless, I hope this paper will accomplish several things. First, it should suggest some of the ways in which behavioral ecology comprises a major conception of person–environment relations. Second, it should indicate some of the ways in which behavioral ecology can guide and influence research on persons in institutional settings. Third, it should present a major model or paradigm of research on health-care environments and other institutional settings.

BEHAVIORAL ECOLOGY

People everywhere are worried about behavioral problems. When they turn to psychology and its allied disciplines for assistance, they confront an irony. Most research in psychology deals with relatively small, discrete problems and with modest objectives and much of the research is locked into certain restrictive methodologies. When one tries to stretch these bits to fit a comprehensive framework and use them to elucidate everyday problems of behavior, the bits and pieces wear poorly. They also wear poorly for those persons who try to use the understanding in a treatment or technology to solve an individual or community problem, and for planners and designers who strain to use the models in designing everyday systems for human living. Human behavior—what persons *do*—is inextricably bound up in the larger conditions of human functioning. By and large, the fragments of our research domain have not begun to approach the scale and complexity of those systems. Simple ideas and simple findings seem to be easier for us to comprehend and embrace than complex ones that may be much more appropriate. Psychology and its allied disciplines must overcome their sense of futility about understanding and changing the seemingly pell-mell course of problems of human behavior and must develop the means to influence and shape that course in more reflective, purposeful, and humane directions. Otherwise, their venerable traditions may disappear and ". . .become like those of a buggy whip factory, interesting and nostalgic but totally useless" (Thoreson, Krauskopf, McAleer, and Wenger, 1972, p. 143). A necessary condition for these developments is the mobilization of a willingness to change our old modes of thinking about human behavior in favor of an approach that accepts and understands the complexity and interdependency of the ecological systems in which behavior is embedded.

Ecology has its origin in the biological sciences and it refers traditionally to the study of relationships between living things and their habitats and the formalization of the natural rules by which those interactions are governed. Complex interdependencies that implicate organisms and their environments in exchanges of energy over time are the scientific coinage of the ecologist and dramatic advances have been made in understanding such interdependencies in areas outside of psychology (Benarde, 1970; Chase, 1971; Colinvaux, 1973; Dubos, 1965, 1968; Margalef, 1968; McHale, 1970; Shepard and McKinley, 1969; Smith, 1966; Wallace, 1972; Watt, 1966, 1968).

These advances are important, but the formulation of the ecological perspective is still very spotty. This spottiness is most evident in the arena of human *behavior*. Much is known about the ways in which man's biological functioning and well-being are caught up in the complex webs of the environments he inhabits. However, very little is known about the laws of human use of environmental systems, the interrelations of those systems with behavior, and the paths by which behavioral problems emerge in those systems. Here is the most pressing need in ecological science.

As just one exemplar of these needs, consider ecological problems of community mental health or community psychology. Professionals in this area operate on at least two basic assumptions: (1) that human development and performance are shaped profoundly by the settings and institutions people inhabit; and (2) that it is legitimate for mental health professionals to promote sound development and prevent maladaptive development by influencing such settings and institutions (Cowen, 1973). However, if we simply ask the question, "How do settings and institutions affect human development?", then we immediately see the dim outlines of a large domain of needed research in social and behavioral ecology. We need systematic frameworks for describing and classifying settings, institutions, and their human involvements. We need to understand the mechanisms and processes by which they function as well as the principles of organization and articulation they display. We need to understand the ways in which behavior is linked to them. And, we need to develop new methods for extensive and long-term monitoring so that we can assess their effectiveness.

In recent years, an ecological approach to human behavior has been urged with increasing frequency and urgency. Among the central theses of behavioral ecology are:

1. Human behavior must be viewed and studied at levels of complexity that are quite atypical in behavioral science.
2. The complexity lies in systems of relationships linking person, behavior, social environment, and physical environment.
3. Such systems cannot be understood piecemeal.
4. Such behavior–environment systems have important properties that change, unfold, and become clear only over long periods of time.
5. Tampering with any part of such a system will probably affect the other parts and alter the whole.
6. We must develop an ecological awareness of the many ways in which simple intrusions can produce unintended effects and the many ways in which long-term harm may follow from short-term good.
7. The focal challenge is to achieve enough understanding of such systems so that the effects of interventions and planned changes can be anticipated in comprehensive fashion.

In other contexts, the implications of behavioral ecology have been discussed for research and theory in psychology (Barker, 1965, 1968, 1969; Moos, 1973, 1974; Willems, 1973a, 1976a, 1976b), technologies of behavior change (Willems, 1973b, 1974), humanistic approaches to behavior (Alexander, Dreher, and Willems, 1976), environmental design (Willems, 1973c, 1973d; Wicker, 1972), and problems of health status and health care (Insel and Moos, 1974; Moss, 1973; Willems and Campbell, 1975).

Behavioral ecology is difficult to define with precision. In fact, it probably will never comprise a single theory or a single set of methods. At its core, it is more on the order of a general view of nature, an orientation, or a set of beliefs, values, assumptions, and principles that leads one to approach the issues of human behavior in distinctive ways. Despite its metatheoretical flavor, behavioral ecology has implications for the construction of theories. It also has profound implications for the ways in which behavioral scientists conduct their work. Another characteristic of behavioral ecology is that it bears on a wide range of issues and problems. At one extreme are problems on a very large scale, i.e., macroecology. An example would be the cultural and behavioral disruptions that follow when a government removes a people from its typical habitat (Turnbull, 1972). At the other extreme are microecological problems; e.g., a child's behavior on the playground deteriorates after his or her behavior in the classroom has been changed through behavior modification. The program of research discussed throughout the paper falls toward the microecological end of this spectrum.

THE PROBLEM OF SPINAL CORD INJURY

During each day and within the life span of the individual, various processes of adaptation lead to the development of a repertoire of behavioral performances in the everyday environment. This repertoire is part of the substance of everyday life and is critical to survival, but it is usually overlooked by the healthy person because it is taken for granted. Onset of a severe physical disability such as spinal cord injury eliminates or alters this repertoire and the person's major means of coming to terms with his everyday enviroment is impaired, often drastically.

Just as the everyday repertoire of performances is taken for granted by the healthy person, so does he take for granted many features of his environment. With the onset of a severe physical disability, the person's environment suddenly becomes formidable, intimidating, and in some cases, insurmountable. Thus, the central problems of rehabilitation cluster around performance and environment. Rehabilitation comprises programmatic arrangements designed to restore or substitute as much as possible in a person's lost or altered repertoire, to teach him new forms of performance and new kinds of relations to the environment, and to alter environments in appropriate ways. The goals of rehabilitation are usually stated in performance or behavioral terms, e.g., "to help the person become maximally independent." "Independence" is a behavioral issue because it points to performances that the person can carry out in his usual environment with a minimum of intervention and support from others. In these terms, rehabilitation means intensive and goal-oriented addition, rearrangement, and substitution in the client's repertoire of behavior and behavior–environment relations.

In addition to being heavily oriented toward functional performance, the perspective and focus of rehabilitation must also be longitudinal. The evolution of the disabled person's repertoire of performance often occurs gradually over time, but the transition to resettlement in the world outside the hospital is not one event; it is also a gradual process and sometimes a lengthy one. Much of the person's time and effort is devoted to carrying out functions and procedures he has been taught, and his well-being depends upon his ability to apply and practice what he has been taught. Thus, the rehabilitation process is a very long one, extending for months or years. Given the importance of what the disabled person does and uses in his everyday environment over a long period of time, it follows that assessment of his status at any given time and his progress or regression over time should focus on his actual performance.

While in the hospital, the patient's behavior is like a continuous stream that sometimes damps down to a minimal level and sometimes quickens to a very brisk pace, sometimes widens or even splits into several simultaneous occurrences and sometimes narrows to one, sometimes intertwines with the behavior of others and sometimes moves in isolation, sometimes changes form or direction at the patient's initiative and sometimes at the instigation of others. Each of the discrete events in that stream occurs at some unique place and time, and yet the stream as a whole flows in a continuous, sequential manner over time. The interface between the patient's behavior stream and the hospital's delivery system is fundamental to the understanding of a hospital. The planners and the agents of a hospital's delivery system assume that their environmental provisions and their role performances are intimately intertwined with patient behavior; if this were not so, there would be no reason to create systems of rehabilitative care. This means that a careful look at patient behavior and its environmental linkages will tell us a great deal about how a hospital system is functioning; i.e., patient behavior is an important criterion of hospital performance and effectiveness.

> Regardless of the method used or the time spent, the single element which seems to have received the least consideration to date in models of the health-care system is the patient. A model which is truly descriptive must include the patient—preferably in terms of his response to behavior of the system. (Pierce, 1969, p. 1700)

Rehabilitation of persons with spinal cord injuries represents a fruitful arena within which to study the microecology of human behavior, for several reasons. First, the aftermath of spinal cord injury is an intensive microcosm of the ontogeny of behavior–environment relations. In a very real sense, the spinal cord injured person is thrust back to an earlier developmental point because of a catastrophic loss. Second, since persons with spinal cord injuries spend relatively long periods in a hospital setting, working out problems of performance and adaptation within the rubric of goal-oriented programs, the period of hospitalization is a period of densely packed and evolving behavior–environment relations. Third, since the conceptions of behavioral ecology relate so closely to the idealisms of rehabilitation and since so little has been known about the development of performance in relation to the environment in persons with spinal cord injuries, the rehabilitation profession needs research of this type.

THE RESEARCH SETTING

At a number of points in the paper, I will refer to a specific, ongoing program of research. In this section, I shall highlight some of the context, methods, and background of that program, which is located at the Texas Institute for Rehabilitation and Research (TIRR) in Houston, Texas. Adults with spinal cord injuries (quadriplegics and paraplegics) comprise one of TIRR's major patient populations. TIRR provides the full range of services—surgery, nursing, physical therapy, occupational therapy, social work, vocational counseling, and outpatient clinics—and it enjoys a worldwide reputation as an excellent rehabilitation hospital.

BEHAVIOR SETTING SURVEY

It is obvious that hopes, plans, and ideas regarding rehabilitation must be conceived by persons or groups of persons and that these plans and ideas must be updated and modified with some regularity. However, it should be just as obvious that the actual delivery of rehabilitation does not live at the level of ideas; it lives and occurs at the level of real and organized arenas of activity and standard programmatic provisions mobilized by the hospital system. Second, the receiving of rehabilitation by spinal cord patients does not occur at the level of hopes and ideas; it consists of making sure that real patients become exposed to as many of the hospital's arenas of activity as possible in a timely and appropriate fashion. It is at these two levels—the hospital's meaningful units of activity and the patient's involvement in the activities occurring there—and the linkages between the two that effective rehabilitation does or does not come about. If that is so, then evaluation of the delivery of care and assessment of patient performance must be addressed to both of those levels. In order to do that, the standard structure of the hospital environment in terms of activity arenas must be identified, described, and measured in a way that is relevant to the delivering and receiving of rehabilitation.

It is for such reasons that we have used the *behavior setting* unit and two behavior settings surveys (Barker, 1968; Barker and Gump,

1964; LeCompte, 1972). The behavior setting is defined and measured in terms of location, physical environment, and patterns of behavior by occupants and its demarcation is based on the degree of interdependence of its component parts and its degree of independence of other settings. The behavior setting unit takes location and physical environment into account, but it is a much more "live" and behaviorally relevant unit than purely architectural units would be. For example, one of the behavior settings at TIRR (a ward setting) includes six architectural divisions (three bed areas, nurses' station, the end of a corridor, one bathroom). Yet, it is *one* behavior setting because the standing patterns of behavior and the flow of persons are common across the six architectural divisions and change beyond them. The importance of this to rehabilitation is that when we measure the diversity of the environment to which a patient is exposed, this whole ward area is counted as one setting and the patient is not counted as having entered a new setting (a new arena of activity) until he actually leaves this whole complex and enters the main corridor, which has its own patterns of behavior.

A behavior setting consists of a combination of social behavior patterns embedded within a physical and temporal structure. The behavior setting unit has the following defining attributes: (1) a recurrent pattern of behavior and (2) a particular physical structure, occurring at (3) a specific time and place, with (4) a congruent relation between behavior and the physical structure. Both the behavioral and the physical components of behavior settings are necessary to their identification. The basic method of the behavior setting survey is to specify a place (the treatment environment of TIRR in our case) and to document all the specific settings that occur there during a delimited span of time (a year in our case). Some occur once and some occur more often. The growing list of possible settings is culled and reorganized by systematic techniques of rating and comparison, and the result at the end of the year is a list of behavior settings, each of which has a consistent, quantitative level of independence from the others (Barker, 1968). Then, by observation, through the use of informants, or by perusal of data such as minutes and work logs, investigators tabulate who actually participates in the settings, broken down into classifications by age, sex, professional category, etc., as well as what various persons do and how long they do it. Standing patterns of behavior (modal behaviors) and mechanisms of behavior (how the things get done) are also rated. The end product is an organized and differentiated picture of how the hospital subdivides itself into functional ecological–behavioral units. In order to avoid biasing factors in data collection, such as those that are often caused by staff turnover and seasonal fluctuation, a relatively long period of time is usually taken as a base period for a behavior setting survey. At TIRR, an entire 12-month year was taken for each of the two behavior setting surveys. In the first survey, the period from 1 July 1968 to June 1969, was used. In the second survey, the same period in 1971 and 1972 was used. These are called the 1968 and 1971 surveys, respectively. We have found that once a setting survey and a setting list have been completed, only minor updating is necessary to keep the list in usable form. Once the list is completed, numerical codes for settings become a part of the system for performance assessment described below.

DEVELOPMENTAL WORK ON PATIENT OBSERVATION

In order to develop procedures for direct monitoring of patient behavior, we conducted programs of observation of patients in 1968 and 1971. We observed 12 patients at TIRR in the summer of 1968 and 15 patients in the summer of 1971, each for one full 18-hour day (from 5:00 A.M. until 11:00 P.M.). In both studies, the groups included all of the spinal cord injured patients who were involved in the hospital's program of comprehensive rehabilitation and included mixtures of ages, sex, races, and variations from early in treatment to predischarge.

A team of trained observers followed each patient continuously for 18 hours by rotating in 2-hour observational shifts. Using a small, battery-operated cassette recorder, the observer dictated a continuous narrative description of the target patient's behavior and enough of the immediate situation to add intelligibility to the narrative. Few strictures were placed on the observational process, but observers were instructed to dictate clock time into the narratives in terms of minutes and fractions of minutes. After typists had transcribed the narratives, observers proofed their own transcripts and one editor screened all of the transcripts for grammar, clarity, and consistency of style. By this process, we obtained 12 18-hour protocols totaling 216 hours of patient time in 1968 and 15 18-hour protocols totaling 270 hours in 1971.

Our analysis of the protocols (Willems and Vineberg, 1970) assumes that they capture and describe the sequential behavior stream of a patient and that the events in that behavior stream include things the patient did, things that were done to him or with him, and periods during which the patient was idle or passive. Our major coding unit, a *chunk*, demarcates a molar event in the behavior stream of a patient that (1) can be readily characterized by a single principal activity, (2) begins at a clearly described starting point, (3) occurs over time in a characteristic, sustained fashion, with all its essential accompaniments, and (4) ends at a clearly described stopping point. The analysis is designed to minimize the coder's inferential burden and to provide the means to retrieve systematic and quantitative information regarding the common, everyday behaviors of patients. Direct involvement by the patient is a necessary condition for marking a chunk. Thus, if the patient is clearly described as having been involved in two distinct principal activities at the same time (for instance, eating lunch and playing cards), two chunks are demarcated as having occurred simultaneously. Examples of chunks are "Watching TV," "Eating Scheduled Meal," "Conversing with Physician," "Passive Range of Motion—Arms," "Waiting," "Transferring," and "Reading Magazine."

Chunks are marked on the protocols by means of marginal brackets. The information on each chunk is then transferred to punch cards, with the following eight codes: (1) patient identifying number; (2) where the chunk occurred, in terms of behavior setting; (3) starting time; (4) ending time; (5) numerical code for the kind of behavior; (6) who else, if anyone, was directly involved in the principal activity of the chunk, and how many persons were involved; (7) who instigated or initiated the chunk; and (8) the degree of involvement by the target patient in the principal activity, on a scale including active, passive, and resistive participation. Thus, for each chunk, it is possible to retrieve information regarding *who* did it, *what* kind of behavior it was, *where* it occurred, *when* it occurred, for *how long* it occurred, *who else* was directly involved in it or whether the patient did it alone, who *instigated* it, and *how actively* the patient was involved in its execution. When these measures are taken singly and in combination and when the overall distributions of chunks are considered, a great many analyses and descriptive statistics become possible.

Longitudinal Observations: Assessment of Patient Performance

Since 1973, we have been observing patients longitudinally as a step toward the goal of developing a longitudinal functional assessment system. In this phase, we gather behavioral data from patients throughout their hospital stays (an average of three months) and for 12 months beyond discharge, for a total period of about 15 months per patient. Although the 15-month period represents a continuous data-gathering period, it can be presented best in terms of inhospital and posthospital phases and their various components.

Inhospital Phase

Observational Measures. The inhospital phase of observation starts on the day a patient is admitted to TIRR. After fully informed and signed consent is obtained from the patient, trained personnel begin to observe him directly on a prearranged schedule. Ten 1-½ hour observations are spaced across each week in such a way that all of the 15 hours between 7:00 A.M. until 10:00 P.M. are observed during the week. Typed transcripts of the observations are coded with a system that was modified from our earlier version on the basis of reliability assessments (Willems and Crowley, 1976).

Instrumented Measures. We have begun to test the simultaneous acquisition of two types of instrument-based measures. The first is gathered by means of a series of pressure sensitive pads that are placed under the patient's mattress and connected to a continuous strip chart recorder in one of the hospital laboratories. This device provides a sensitive, continuous recording of motility on the bed surface and, more importantly, an exact measure of time out of bed. The second is gathered by means of a set of mechanical odometers attached to the two wheels of the patient's wheelchair (Alexander, 1976). These odometers provide a measure of patient mobility. We have tested the reliability of the odometers and the adequacy with which periodic readings estimate total mobility. Data from the bed monitor and odometers are then mapped onto the performance data to ascertain which observational measures are approximated best, or estimated best, by the instrumented measures. These efforts are motivated by the potential saving in time and effort that would come from using instrumented measures as much as possible to supplement or replace the more cumbersome observational measures.

Posthospital Phase

Activity Record. Two weeks before a patient-subject is discharged, he is trained to keep a standard *activity record* (Widmer, 1976). Either on a standard form or on a cassette recorder (for patients who cannot write), the subject keeps a sequential account of performances or activities (much like chunks) and provides the following information for each activity: beginning time; location; the label of the activity;

whether assisted and if so, by whom; how transport was executed if it occurred; whether conversational interaction occurred; and other companions. The activity records and observations yield systematic data on the number of activities engaged in, the number of settings entered, the rate at which activities are conducted alone or with help, etc. In addition to studying such measures for fluctuation and change, we relate them to the behavioral measures obtained while the patients were in the hospital. Gathering of posthospital activity records begins when the patient-subject is discharged. Each expatient produces one complete day's activity record every 10 days (from waking up until going to bed). The activity records are supplemented by occasional observations by the research team and by interviews with the expatient and the person primarily responsible for his care.

Environmental Survey. Once every month, a member of the research team visits the expatient's home to conduct a survey of the home environment (Baker, 1976). The first step is to construct an inventory of elements of the expatient's home environment that are potentially negotiable from the vantage point of a nondisabled person; e.g., switches, doors, closets, drawers, appliances. Second, with the expatient actually attempting to negotiate each element, an observer then rates each element in the inventory for whether or not the expatient is able to negotiate it by himself (with only his ordinary assistive equipment).

We conduct these pilot environmental surveys for several reasons. First, we wish to understand better the kind of behavior–environment relations involved in life outside the hospital. Second, the survey provides an objective, behaviorally based assessment of the overall negotiability rate of the person's environment and clear documentation of the areas and subareas that are not negotiable for him. Third, we find from the activity records that the expatients' levels of independent performance change over time. Documentation of changes that have been made in the home environment should help us to understand the changes in independence. Finally, in the longer range, we are also committed to contributing to the habitability of expatients' environments. The systematic, objective, client-based approach to environmental assessment embodied in the environmental survey is one major step in the direction of potential counseling for environmental modification.

KEY ASPECTS OF BEHAVIORAL ECOLOGY

In this section, I will present eight major aspects or guiding principles of behavioral ecology. In the case of each, there will be a general presentation and some background, followed by some illustrations from our research program. The illustrations will be in the form of findings, generalizations, or characteristics of the program.

MOLAR PHENOMENA AND NONREDUCTIONISM

The ecological perspective tends, generally, to place more emphasis upon *molar* phenomena than upon *molecular* ones. Closely related is a relative emphasis upon environmental, behavioral, and organismic holism and simultaneous, complex relationships. This is so in part because, all the way from survival of a species, through adaptive functioning, down to day-to-day and moment-by-moment adaptive processes, the emphasis is upon the organism's and the population's behavioral commerce with the environmental packages they inhabit. Adaptation to everyday settings and long-range, functional performance in them places focus on coming to terms with the environment, on what Powers (1973) calls *results* rather than more molecular movements. Even though it might be possible, in principle, to study the ecology of eye blinks (a relatively molecular phenomenon), the behavioral ecologist usually focuses on larger, setting-sized, functional behavior episodes; e.g., cooperation, conflict, solving problems, social interactions, transporting, etc. (Barker, 1963a; 1965; 1969; Sells, 1969; Wicker, 1972; Wright, 1967).

In some disciplines, such as neurophysiology, molecular movements (e.g., flexion or extension of a limb) can be important. For example, where the primary question is the mechanism of neural control, standard molecular movements can provide a medium for testing neural control. In an ecological perspective on behavior, there is much more concern about the adaptive match between performance and environment, about the separate and reciprocal effects of behavior and environment upon each other, and about the linkages and rhythmicities that characterize behavior–environment systems. When one focuses on effects and adaptive results in behavior–environment relations, one is almost inevitably focusing on molar phenomena. That is, clusters of movements are identified and labeled in terms of their adaptive values, meanings, and results *as clusters*. Ecologically, it is both more efficient and more appropriate theoretically to speak of "putting trash out," "driving car to work," and "brushing teeth," rather than a series of discrete movements for each case.

In our program of research, we are primarily concerned with molar units in two domains, the first of which is the environmental side. Here, we begin with behavior setting units and shift to a more molecular level only when a more molecular aspect of the environment

becomes salient; e.g., when a bed is located in such a way that the
patient cannot transfer to his wheelchair. In the second domain,
patient behavior, we focus on units of performance that have meaning
in everyday life. The *chunks* in our data system represent the kinds of
molar performances that are crucial to everyday functioning; e.g.,
eating a meal, transferring from wheelchair to bed, grooming, reading
a book, transporting from ward to physical therapy, working on a
computer program. Patient performances of this sort vary in terms of
frequency, diversity, length, location, and the manner in which they
are executed; e.g., from docile and dependent (aided) to active and
independent (unaided). We have chosen this molar focus because
independent functioning occurs only when such *molar* acts are executed alone and at one's own initiative. Since it is these functional
endpoints that count most in effective rehabilitation, the hospital
system should understand the processes that affect the acquisition of a
repertoire of such molar acts.

BEHAVIORAL FOCUS

Traditionally, psychologists have studied many aspects of human
functioning: thoughts, feelings, moods, structures of personality,
habits, motives, judgments, cognitions, self-concepts, beliefs, attitudes, perceptions, sensory processes, ego strengths, and a host of
other traits and dispositions. The anchoring point for the behavioral
ecologist is individual and collective behavior, the molar performances
of persons vis-à-vis the environment and their molar responses to it,
e.g., where persons go, what they do with their time, how they use
resources, how they interact with each other. This emphasis has
emerged for several reasons. First, the great strides made in animal
ecology and ethology have had an influence. In those areas, investigators have been forced to focus on overt behavior because the subjects
cannot report feelings and subjective states.

Second, not only do attempts to measure subjective and disposi-
tional variables often evolve into endless psychometric jungles of
checking, cross-checking, and validation, but, more distressingly,
they often yield very weak correlations to behavior (Mischel, 1968;
Wicker, 1969, 1971; Willems, 1967). On the basis of his review of the
attitude–behavior field, Wicker (1969) concludes that there is

> . . .little evidence to support the postulated existence of stable, underlying
> attitudes within the individual which influence both his verbal expressions
> and his actions. . . .Most socially significant questions involve overt
> behavior, rather than people's feelings, and the assumption that feelings
> are directly translated into actions has not been demonstrated (p. 75).

Third, with a shift in emphasis to problems of everyday life and to naturalistic research, psychologists have begun to question some old, dearly held assumptions, one of which asserts that affective, attitudinal, and cognitive variables affect behavioral variables in a simple and unidirectional way. Bakeman and Helmreich (1975) argue:

> Much of the laboratory research of the last two decades in fact has viewed cognitive variables as affecting behavioral ones. Perhaps it is worthwhile to speculate why this is so. The experimentalist. . . manipulates what he can most effectively manipulate, and cognitions are far easier to manipulate than actual behavior. Perhaps feasibility comes to bias views of causality, and then theorists working from laboratory data are likely to become convinced that cognitive variables determine behavioral ones. (p. 488)

Fourth, behavioral performance and response are the major means by which person–environment adaptations are actually mediated. The ecologist does not argue against the reality of subjective phenomena. Rather, his position is a matter of emphasis in seeking to understand human functioning in relation to the environment. Nor does the relative emphasis on behavior imply a simplistic advocacy of behaviorism. To the ecologist, overt behavior simply is more important than many other psychological phenomena. For the ecologist, it is more important to know how parents *treat* their children than how they feel about being parents; more important to observe whether or not passersby *help* someone in need than what their beliefs are about altruism and kindness; more important to note that a person *harms* someone else when given an opportunity than to know whether his self-concept is that of a considerate person; more important to know what persons *do* with trash than to measure their attitudes about waste disposal; more important to know what one *does* in the way of consuming alcohol or hiring women than to infer community beliefs regarding alcohol or women. To the behavioral ecologist, person–environment-*behavior* systems represent problems to be understood and solved that are simply more important than person–environment-*cognition* systems or person–environment-*attitude* systems.

Despite these developments, measures of cognitive and affective response to the environment remain popular, or even dominant, in the burgeoning field of environmental psychology (Altman, 1975; Ittelson, Proshansky, Rivlin, and Winkel, 1974). In fact, Ittelson *et al.* assert, "The single human psychological process most critical for man/environment interaction, and the one that underlies all the response characteristics that we have described, is that of cognition" (p. 98). I disagree because behavior, in the sense of *doing* things overtly, is the principal means by which persons make long-range adaptations to the environment and it is the means by which they modify the environment. It is not readily apparent to me how all of the data on how-it-looks, how-it-feels, and what-people-think-they-want will become translated into understanding these problems of long-term environmental adaptation and adjustment (see Danford and Willems, 1975).

We have followed this behavioral bias in our program of research and we have emphasized a behavioral perspective throughout. The philosophy of rehabilitation is often stated in behavioral terms (e.g., "adjustment," "independence"), but the actual assessment of rehabilitation has seldom been approached in quantitative, behavioral terms. For a spinal cord injured person, the progress he displays in what he does, how he does it, and where he does it is the critical issue. The period of hospitalization is a period of highly focused preparation for life outside the hospital. Thus, not only is it important to assess change in the patient's emerging repertoire of adaptive behaviors, but it is important to assess his emerging accommodation to the environment and his interface with it. Stated in comparative terms, it is more important to measure change in his performance over time than to measure changes in his cognitions over time.

NATURALISTIC EMPHASIS

The ecological perspective on behavior emphasizes complex and systemlike regularities and interdependencies among organisms, behaviors, and environments, often over long periods of time. Behavioral ecology is not primarily a methodological orientation, but it has many implications for methodology, as has been suggested here and elsewhere (Barker, 1965, 1968; Willems, 1973a, 1976a; Alexander et al., 1976).

Behavioral ecology is largely naturalistic in its methodological orientation; "largely" because it is not defined by any particular methodology and because this is an emphasis rather than a necessary condition (Willems, 1973a). Pluralism of methods is crucial to behavioral ecology, but the ecologist's methodological statement of faith has two parts. First, with Keller and Marian Breland (1966), the ecologist says, ". . .you cannot understand the behavior of the animal in the laboratory unless you understand his behavior in the wild" (p. 20). Second, contrary to widely held canons in psychology, the ecologist believes that the investigator should manipulate and control only as much as is absolutely necessary to answer his questions clearly, an argument that has been made by many writers (Barker, 1965, 1969;

Brandt, 1972; Chapanis, 1967; Gump and Kounin, 1959–1960; Menzel, 1969; Willems, 1965, 1969). The ecologist works with the continual reminder that holding experimental conditions constant while varying a limited phenomenon is a figment of the experimental laboratory that may result in the untimely attenuation of both findings and theories. The ecologist recommends more dependence on direct, sustained, naturalistic observation of human behavior and less on shortcut methods based upon verbal expression and the handiest investigative location, which so often is the experimental laboratory. Questionnaires, interviews, tests, and experiments all have one important characteristic in common: They require the subject to interrupt what he is doing in his natural context and perform a special task for the investigator.

In our program of research, we have aspired to use naturalistic methods, i.e., techniques for data gathering that intrude minimally on the usual, ongoing behavior of the patient and the hospital. One of the most informative ways to document a patient's behavior and to learn the relation between what he does and the context within which he does it is to follow the patient firsthand and generate a continuous, systematic account of his behavior stream and the immediate situations through which he passes and with which he interacts. The theme is the development of methods to monitor the ongoing behavior of patients during rehabilitation. Such methods are preferable to the more traditional technique of removing the patient from his usual sequence of activities to participate in a special interview or test of some sort. We prefer to get at the typical behavior directly, as it occurs in its usual manner, because that is where preparation for posthospital life takes place.

Sustaining this program means that the system of naturalistic observation must be maintained for long periods of time. For example, one recent patient was observed over a period of 140 days (from admission to discharge), for a total of 300 hours. Many persons express concern about the reliability of this kind of data system (e.g., Bijou, Peterson, and Ault, 1968; Johnson and Bolstad, 1973). With our specimen record system, it is quite routine to measure intercoder agreement (e.g., two independent analyses of one observer's protocol). However, assessment of interobserver agreement raises some problems because the direct product of an observer's work is an uncoded, narrative protocol. We can report our solution to this problem and some illustrative data on reliability from the 13-week hospitalization period of one patient (Dreher, 1975).

During one 1½ hour observational session each week (10% of the data), two observers produced observational protocols on the patient simultaneously. After typing and photocopying, the protocols were coded in the usual way, but according to the scheme outlined in Figure 1. Two separate persons (A and C) coded the protocol produced by one observer, but they did the coding independently. Two more persons (B and D) coded the protocol produced by the second observer. Comparisons were then made in terms of the percent of total protocol material that was accounted for in exactly the same way. Comparing A-1 to C-1 and B-2 to D-2 gave two estimates of intercoder agreement each week, i.e., *intercoder reliability*. Next, comparing A-1 to B-2 and C-1 to D-2 gave two estimates each week in which both differences between coders and differences between observers could appear. Since the latter was a comparison between two completely independent processes through the data system, we called it intersystem reliability. Intercoder reliability yields a direct estimate of error attributable to coders. Subtracting intersystem reliability from intercoder reliability yields an estimate of error attributable to observers. Table 1 displays the two reliabilities obtained for each of 13 weeks

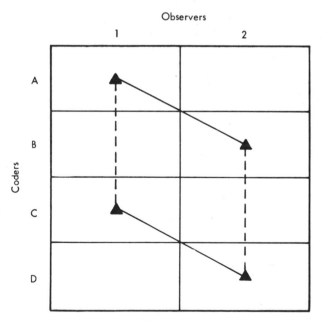

Figure 1. Arrangement by which four coders were assigned to protocols produced from simultaneous observations by two observers. (▲——▲) Intersystem comparison, (▲– – – ▲) intercoder system.

TABLE 1
PERCENT RELIABILITY OF TWO TYPES FOR
THIRTEEN WEEKS

Week	Intercoder reliability	Intersystem reliability
1	99.85	99.14
2	99.50	96.26
3	97.53	93.98
4	99.30	97.90
5	99.20	91.90
6	99.44	96.84
7	98.17	90.20
8	97.40	93.15
9	100.00	94.39
10	96.15	88.21
11	98.36	90.85
12	99.62	95.29
13	99.00	97.98
Mean	98.75	94.31

from the test subject (19.5 hours of reliability assessment). Several things stand out. First, the reliability levels are high and compare very favorably to reliabilities of other observational procedures. Second, coding problems account for small error rates, on the average. Observer problems, though accounting for slightly higher error rates, are still small. Further analyses showed that there is no systematic fatigue effect on reliability (comparison of first, middle, and last portions of observational periods), no systematic warm-up effect (comparison of first, second, and third 10-minute segments), and no systematic effect for time of day (comparison of four quarters of the observational day).

For an observational monitoring system that must be maintained for very long periods of time, the next important question is whether the system and its human components change over long time periods; i.e., is the system stable over time in rendering information? To answer this question, we filmed 60 minutes of a patient's behavior on videotape. Eight observers performed the standard type of protocol observation once and then again 90 days later. All resulting protocols were coded according to the usual procedures. Under these conditions, with observed material held constant and the same observers working 90 days apart, there were negligible differences in agreement rates among (1) the same observers and coders across time, (2) different observers and coders at one time, and (3) different observers and coders across time. In other words, the data system functioned at a high level of stability across time.

Analogous tests of reliability have been conducted on the activity records and environmental surveys for expatients. Reliabilities are high. What is reliability to the researcher is quality or dependability to the practitioner. Since hospital personnel have begun to use our data in making assessments and treatment decisions, we now check reliability routinely on 10% of all data collected.

DISTRIBUTION OF PHENOMENA

Behavioral ecology concerns itself with documenting the distribution of phenomena in nature; that is, the range, intensity, and frequency of behavior–environment relations in the everyday, investigator-free environment. Somehow, this basic issue is poorly understood by psychologists and such research is not yet widely accepted, valued, or promoted (Elms, 1975; McGuire, 1969; Willems, 1969). Perhaps this is so because the model of experimental and analytical science and the development of technology is so appealing. Not only are distributional aspects of behavior and behavior–environment relations important to basic ecological understanding, but if we are ever to achieve a comprehensive approach to influencing, changing, and accommodating human behavior, then we must begin by taking account of (1) the much larger organism–environment–behavior systems within which such approaches take place, (2) distributions within those systems, and (3) optimal proportions and combinations of behaviors across populations and subpopulations.

Despite the relatively low status of such research, significant steps have been taken. Some important anthropological, sociological, and ethological studies have gathered distributional data. Closer to a direct focus on human behavior, Ebbesen and Haney (1973) made more than 13,600 observations of drivers at intersections and the environmental conditions under which the drivers took the risk of turning in front of oncoming traffic. The resulting tables of frequencies are full of implications for understanding the everyday behaviors of drivers and the factors that affect such risk taking. The pioneering descriptive work of Barker and Wright and their associates on behaviors of persons in their home communities is so rich and diverse that it defies a brief characterization (see Barker, 1963b, 1965, 1968; Barker and Schoggen, 1973; Barker and Wright, 1955; Wright, 1967, 1969–1970).

In addition to providing descriptive characterizations of ecological systems, distributional data can also suggest where scientific straw men are. Many early theories of child development placed singular importance on the concept of frustration. Fawl (1963) analyzed records representing over 200 hours of observation of children. Occurrences of blocked goals (frustration) were relatively few. When they did occur, the child usually did not appear disturbed, and the few disturbances were of a mild intensity. In another area, it has often been assumed that disfigurement of the face produces widespread negative effects in a person's life because of social stigma surrounding the face as a crucial social stimulus. After 56 days of monitoring the everyday social and behavioral lives of children and adolescents with grotesque facial disfigurements resulting from burns (as well as matched control children), Schmitt (1971) and Ronnebeck (1972) found that facial disfigurement produced almost no disruptions of the subjects' daily social lives. Stuart (1973) conducted a similar study of wheelchair-bound college students for seven days and found that the overall frequencies of various types of behavior, participation in various settings, and time spent in various activities and places for wheel-chair-bound students were almost indistinguishable from the behavioral topographies of matched, nondisabled students.

In our program of rehabilitation research, we have observed some 45 patients for a total of about 2,000 hours to date. In our current phase, in which we observe patients regularly from admission until discharge, we have already accumulated 1,600 hours of observational data covering the span of hospitalization. The descriptive distributions suggest that, in some fundamental respects, the treatment environment and the behavior of its inhabitants display astonishing stability over time. For example, between 1968 and 1971, the number of behavior settings in the treatment areas of the hospital only changed from 122 to 120 and the frequencies of various kinds of settings remained very constant. The distribution of patient behavior, across both behavior settings and categories of behaviors, remained very stable between 1968 and 1971

Some settings (e.g., wards) capture many different kinds of patient behavior while others (e.g., cafeteria) are more sparse, behaviorally. Some patient behaviors (e.g., conversing) occur in almost all settings, while others (e.g., exercising) occur in very few. Patients enter only a fraction (about 17%) of the treatment-related settings, but that fraction is the subset in which most of the occupancy and participation by staff and patients occur. Furthermore, almost 90% of patient behavior takes place in an even smaller subset of the settings (five settings, or about 4% of the hospital's total). Using a behavioral mapping technique in a rehabilitation hospital (i.e., periodic record-ing of patient behavior in various locations rather than continuous observation of one patient at a time), Miller and Keith (1973) con-firmed our findings that (1) patients spend much more time in wards than treatment areas, (2) patients spend more time idle and socializing than in treatment, and (3) treatment-related activities show huge fluctuations during the working day.

The observations yield detailed, quantitative descriptions of rep-ertoires of patient performance; e.g., number of chunks (pace), num-ber of different kinds of chunks (behavioral diversity), number of entries into settings (mobility), number of different kinds of settings entered (environmental diversity), number of instances of overlapping behavior (complexity), rate of self-instigated behaviors, rate of active behaviors, rate of behaviors carried out alone (unaided performance), rate of independent behaviors (self-instigated and unaided), rate of zestful behaviors (self-instigated and active), number of settings in which independence is displayed, and so on. The measures are very sensitive to patient differences and to changes within patients over time. For example, patients display much higher rates of mobility, independence, complexity, zest, and diversity near discharge than they do earlier, near admission. The observations show that patient behavior is highly dependent on location in the hospital; i.e., patients display more independence, zest, and behavioral complexity in some settings than others. We find that the observations allow us to document some of the ways in which patient behavior is responsive to changes in hospital programs. Finally, the data provide systematic descriptions of the ways in which involvements with patients are distributed among staff members and the ways in which staff involve-ments with patients are distributed across behavior settings and kinds of patient behavior.

We are continuing this work and extending it to include system-atic documentation of posthospital adjustment. We have sustained this effort not only to provide the relevant professions a complete distributional documentation of the aftermath of spinal cord injury, but because, having such complete distributional data in hand, we now can do much more than speculate about many important ques-tions in the complex and expensive domain of rehabilitation.

ENVIRONMENT AND BEHAVIOR

In their discussion of techniques for measuring behavior–environ-ment relations, Ittelson *et al.* (1970) present a truism that is also a poorly understood principle of man–environment relations: "Behavior always occurs someplace, within the limits of some physical surroundings" (p. 658). One of the central conceptual issues of behavioral ecology is the transactional character of organism–environment systems (Craik, 1970; Barker, 1963a, 1969; Gump, 1969; Alexander *et al.*, 1976). Studying molar behavior by itself is always shortsighted because behavior points two ways, or relates in two directions—organism and environment—and to mediating processes between them. Behavior represents the coming-to-terms that occurs in organism–environment systems. Organisms and environments are part of the same system, with behavior being the major interface between them. Watts (1969) points out:

> If you will accurately describe what any individual organism is doing, you will take but a few steps before you are also describing what the environment is doing. . .we can do without such expressions as "what the individual is doing" or "what the environment is doing," as if the individual was one thing and the doing another. If we reduce the whole business simply to the process of doing, then the doing, which was called the behavior of the individual, is found to be *at the same time* the doing which is called the behavior of the environment. (p. 140)

Thus, the ecologist would argue, we must study *behavior–environment* units rather than just behavior units. Without that perspective, the bits and pieces which we study so frequently in experiments and with which we tinker so indiscriminately in psychotherapy, behavior modification, and behavioral pharmacology are all abstractions that have lost much of their scientific and practical usefulness because they are separated from the contextual interdependencies in which they occur in everyday life. What is worse, when changing only the persons and such abstracted bits of behavior, we may create unintended effects in the ecological systems with which the behaviors are linked (Rhodes, 1972; Willems, 1974). Finally, when we look to performance for diagnostic judgments and assessments of how well a person is doing, we may be misled if we restrict our judgments to behavior alone rather than behavior–environment units.

Much of the recent thrust of our research on persons with spinal cord injuries has been predicated on these assumptions. For such persons, where restoration of functional performance is the central issue, we believe that diagnostic and prognostic judgments based on behavior rates alone will be less complete, less accurate, and less effective than judgments based on changes in behavior–environment units. We are seeking to develop new, functional performance indices of health status and progress, and we are basing them on conceptions of behavior–environment interdependencies rather than conceptions of behavior only. Several examples will illustrate this emphasis. At the simplest level, we find that the number of entries into hospital settings (simple mobility) and the number of different kinds of settings entered (environmental diversity) are rather sensitive indicators of patient progress over time. After an early period of very limited mobility, it appears (and we hypothesize) that patients who ultimately do well outside the hospital go through a midhospitalization period of very rapid increase in mobility and environmental diversity, after which there appears a leveling-off prior to discharge.

Second, for each unit of behavior, our observational data yield information on what kind of behavior it is, who else is involved with it (whether directly, hands-on, or by means of conversational interaction with the patient), and where it takes place. Thus, for given

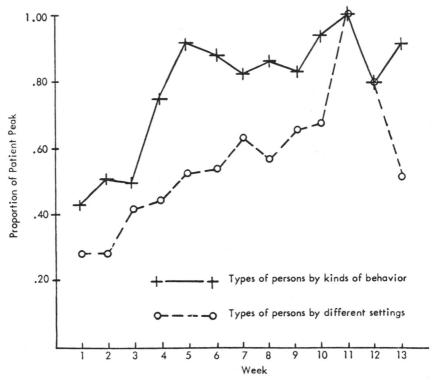

Figure 2. Weekly numbers of (a) combinations of kinds of behavior and types of persons patient was involved with and (b) combinations of behavior settings and types of persons patient was involved with (expressed as proportions of patient's maximum).

periods of time throughout hospitalization, we can calculate (1) the number of different combinations of behavior and type of other person in which social interactions are involved (social–behavioral measure) and (2) the number of different combinations of hospital setting and type of other person in which social interactions are involved (social–spatial measure). Figure 2 displays the weekly number of combinations of type of patient behavior and type of other person interacted with (solid line) and weekly number of combinations of behavior setting and type of other person interacted with (broken line) for one patient. Each weekly measure is shown as a proportion of the patient's maximum, which occurred during week 11. We find that growth of the patient's social–behavioral world begins at a higher level and approaches the maximum much earlier than growth in his social–spatial world. This probably occurs because new combinations of activity and sociability can be produced without leaving a given setting, whereas it is more difficult to produce new combinations of place and sociability, which requires entering new settings. Thus, we hypothesize that change in a patient's social–spatial world will be more *diagnostic* of progress.

Third, an even stricter performance measure is change in the patient's spatial–behavioral world; i.e., the number of new combinations of setting and behavior. Figure 3 shows weekly change in this measure across the hospital stay of two patients.* When complete inhospital and posthospital data are available for a larger sample of patients, we predict that patients displaying inhospital patterns like patient 33 in Figure 3 will show stronger patterns of posthospital performance than patients like number 34, because patients with the former picture will have built up larger repertoires of performance-in-settings. In other analyses, we will tighten the criterion, e.g., testing the predictive power of combinations of settings and behaviors conducted alone (environmental diversity of unaided performance).

Finally, the data yielded by the survey of negotiability of the home environment (see above) are a somewhat different example of the application of the behavior–environment conception. The measure

---

* Figure 3 and several subsequent figures have been plotted by computer directly from our cumulative file. I have used this format for display here because it demonstrates the kind of graphical display of patient performance that we can produce on very short notice for our research staff or for members of the hospital staff. Furthermore, this format illustrates the kind of visual display that members of the hospital staff will soon be able to call up instantaneously on the viewing scopes of various computer terminals in the hospital. Thus, when a staff member wishes to know at any time how a patient is doing in terms of our observational assessment of performance, the information will be available.

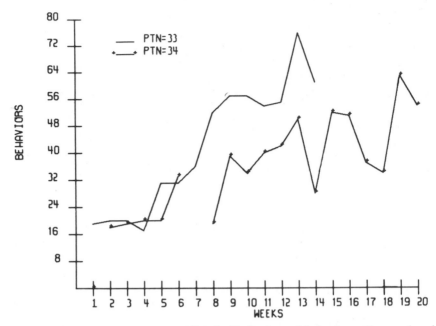

Figure 3. Number of combinations of kind of behavior and behavior setting produced each week by two patients across their periods of hospitalization.

is based on the assumption that the negotiability or manageability of an environment is not simply a function of the person's behavioral capacity *or* the properties of the environment, but is a function of both taken together, i.e., their interface. Thus, the measure reflects the proportion of environmental components in a home that are actually negotiable by the particular person. Figure 4 shows the results of nine such surveys for one expatient in our study sample. This expatient, a quadriplegic, lived in the home of his sister through week 32 postdischarge (first continuous segment of the graph). A few days later, he had moved to a cooperative residential complex that is physically adapted for use by severely disabled adults. Results of three environmental surveys in the new residential setting (after the break in the graph) indicate that, for this person, the new environment is much more negotiable; i.e., he can manage and use a higher proportion of its components by himself. The important point here is that negotiability is not judged for an environment by itself or for a person's capacity by itself, but is based on measuring the functional interface of performance and environment directly.

Our data suggest that frequency or magnitude of behaviors alone

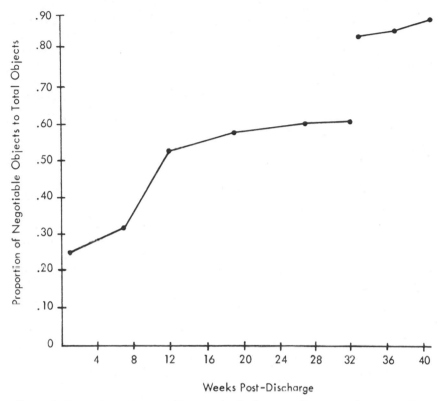

Figure 4. Proportions of negotiable parts in the home environment of one expatient.

are much less powerful indicators of progress and readjustment than change in behavior–environment *combinations*. Thus, the behavior–environment conception of the ecologist is translated into a set of measures that bears directly on problems of health and health status, and behavior–environment units enter into assessments of how well the process of rehabilitation is going.

SITE SPECIFICITY AND PLACE DEPENDENCIES

Closely related to the issue of understanding human functioning in terms of behavior–environment units is the issue of predictability from place, or setting, to behavior. To the behavioral ecologist, *where* organisms are located is never unimportant or accidental because behavior and place concatenate into lawful, functioning systems (Barker, 1963a; Moos, 1973; Wicker, 1972; Willems, 1965). "The correlation between site and activity is often so high that an experienced ecological psychologist can direct a person to a particular site in order to observe an animal exhibiting a given pattern of behavior" (King, 1970, p. 4).

In their discussion of basic assumptions regarding the influence of the physical environment on behavior, Proshansky *et al.* (1970a) argue that observed patterns of molar behavior in response to a physical setting persist regardless of the individuals involved. From their studies of persons in mental hospitals, the same investigators conclude that such intrasetting continuity of behavior often occurs even though inhabitants are cognitively unaware of the structural aspects of the settings. Barker (1963a, 1968) also points out that place–behavior systems have such strong principles of organization and constraint that their standing patterns of behavior remain essentially the same though individuals come and go. Wicker (1972) calls this behavior–environment congruence. Barker (1968) calls it behavior–milieu synomorphy and argues that the appropriate units of analysis for studying such synomorphic relationships are *behavior settings*, whose defining attributes and properties he has spelled out in detail (1963a, 1968).

A clear example is found in the work of Raush and his colleagues in their studies of normal children and children who had been diagnosed as hyperaggressive or disturbed (Raush, 1969; Raush, Dittmann, and Taylor, 1959a, 1959b; Raush, Farbman, and Llewellyn, 1960). By observing the children for extended periods of time in various settings and then examining the frequencies of various kinds of behavior by the children toward peers and adults, the investigators were able to demonstrate several aspects of place dependency. First, the interpersonal behavior of all the children varied strongly from one setting to another. Second, and perhaps most revealing, the place dependence of behavior was much stronger for normal children than for disturbed children; i.e., the influence of the setting was greater for normal children. Finally, as the disturbed children progressed in treatment, the place dependence of their behavior came to approximate the normal children more and more.

Wahler (1975) observed two troubled boys periodically for three years in home and school settings. He found (1) that behaviors clustered differently in home and school settings, (2) that the clusters within each setting were very stable over time, and (3) that different patterns of deviant behaviors occurred in stable fashion in the two settings.

From the evolutionary standpoint, it makes sense to argue that behavioral responsiveness to settings is selected for, because location-appropriateness of behaviors is crucial to adaptation in many settings. The implications of such phenomena are widespread. Two that have become part of the behavioral ecologist's credo are, first, that behavior is largely controlled by the environmental setting in which it occurs and, second, that changing the environmental setting will result in changes in behavior. The third implication is related to methodology. This is the investigative problem of describing and classifying the types and patterns of congruence between behavior and environment and formulating principles that account for the congruence. This effort is important because it promises to contribute much to programs of environmental design. To accomplish this goal, investigators must become more persistent in adding descriptions and codes for locations and context to their measures and descriptions of behavior.

These principles are illustrated and elaborated by our program of research on persons with spinal cord injuries. First, when we look at distributions of patient behavior within different settings, we find that these profiles of behavior vary dramatically from one setting to another (LeCompte and Willems, 1970; Willems, 1972a, 1972b). Some behaviors occurring in one setting do not occur at all in others and the relative frequencies and percentage weights of behavior show strong variation between settings. *What* patients do varies systematically from one setting to another.

Second, in addition to these topographical dependencies on settings, we find dependencies in the more dynamic aspects of patient behavior. From the observations, we extract behavioral measures of *independence* (i.e., the proportion of performances that patients initiate and execute alone) and *zest* (i.e., the proportion of performances that patients initiate and carry out actively). Because increases from very low rates in these measures reflect a relative normalizing of the patients' behavior repertoires, both relate closely to important goals of the hospital's treatment system. Many traditional, person-based theories of human behavior in psychology would assume that independence and zest are largely a matter of individual motivation and thus should reflect a high degree of personal constancy across situations. What we find instead is that behavioral independence varies dramatically when patients move from one hospital setting to another. Table 2 illustrates this variation by settings from day-long observations of 12 patients. Since these wide setting variations are based upon data from all 12 patients, it is possible that the differences were produced by different patients who entered the settings at different rates. To test this alternative hypothesis, we calculated combined indices of inde-

TABLE 2
CHARACTERIZATION OF BEHAVIOR SETTINGS IN
TERMS OF PATIENT INDEPENDENCE

Setting	Rate of patient independence
Cafeteria	.64
Hallways	.48
Outside the building	.30
Stations 1–3 (ward)	.30
Station 4 (ward)	.24
Occupational therapy (OT)	.15
Physical therapy (PT)	.08
Recreational therapy (RT)	.02

pendence for three settings (cafeteria, hallways, wards) and compared them to combined indices for a second set of three settings, (OT, PT, RT) for each of the 12 patients. In the case of each patient, the results corroborated the patterns found above. For every patient, the rate of independence dropped as he moved from cafeteria–hallways–ward to OT-PT-RT.

Third, we find in many cases that differences between settings account for more variance in patient performance than do differences between patients or other variables. Table 3 illustrates this phenomenon descriptively by displaying rates of zest for settings, kinds of patient behavior, types of other persons involved in the behaviors, and individual patients from the same sample of 12. Settings produce more variability than the other factors. It is also interesting to note that the settings and kinds of behaviors that are *accompaniments* of the rehabilitation process (cafeteria, hallways, eating, recreation) produce more zest than those settings and behaviors that are central to rehabilitation (OT, PT, transferring, transporting, exercising).

Fourth, and most interestingly, there are powerful variations among settings in the rate of growth and behavioral development displayed by patients. That is, patients show more goal-oriented change in some hospital settings than in others. To illustrate this phenomenon, Figure 5 displays proportions of behaviors that were carried out actively in three major settings by one patient who was observed longitudinally throughout his hospital stay. On the abscissa, the patient's hospital stay is divided into thirds. Overall, this patient showed great improvement in his rate of active, energetic performance

TABLE 3

CHARACTERIZATIONS OF SETTINGS, CATEGORIES OF BEHAVIOR, INVOLVEMENTS WITH SEVERAL GROUPS OF PERSONS, AND INDIVIDUAL PATIENTS, IN TERMS OF PATIENT ZEST

Settings		Categories of patient behavior		Groups of persons involved with patients		Individual patients	
Cafeteria	.83	Eating	.68	Other patients	.60	Pt. 5	.52
Hallways	.52	Active recreational	.63	OT	.51	Pt. 12	.43
OT	.47	Conversing	.56	Nurses	.33	Pt. 4	.42
RT	.44	Transporting	.50	PT	.32	Pt. 10	.41
Outside	.36	Transferring	.39	Physician	.29	Pt. 6	.39
PT	.29	Exercise and perf. training	.31	Aides and orderlies	.28	Pt. 2	.32
Ward 4	.29	Passive recreational	.24			Pt. 8	.24
Wards 1–3	.22	Nursing care and hygiene	.22			Pt. 11	.24
						Pt. 1	.20
						Pt. 3	.20
						Pt. 7	.18
						Pt. 9	.17
Range	.61		.46		.32		.35

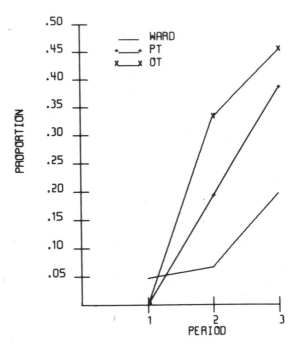

Figure 5. Proportion of active performances to total performances by one patient in three different hospital settings.

(as against docile and relatively passive performance). However, the degree to which he displayed that progress depended strongly on where he was located in the hospital. This patient showed three times as much progress in OT as he did in the ward. We find analogous setting dependencies in the cases of all patients and we find them for the various aspects of performance, i.e., rate of self-instigation, rate of performing along (unaided).

In summary, not only do we find that persons perform differently in different settings when we make simple comparisons between settings, but persons change in different ways and at different rates in different settings. When these central principles of behavioral ecology really soak in, they will affect human behavioral science in profound ways. One area that will be affected strongly is the area of human assessment. It will no longer be so tenable or defensible to assess *a person's* performance. Rather, we will have to assess performance-by-settings, simply because variations in settings produce variations in performance.

Closely related is the clear hint that many traditional approaches to diagnosis—judgments of what is wrong or how the client is doing—are far too narrowly defined, facile, and restricted. A case in point occurred when we began participating in patients' treatment teams in our program of research. Often, a staff member representing one service (e.g., nursing) at patient rounds would offer a judgment of how the patient was doing that would be highly discrepant with the judgment of a staff member representing another service (e.g., occupational therapy). Each assumed that he or she was giving a valid judgment of *the patient*. The interesting point here is that our data suggested in many cases that each staff member was responding to the patient's performance in a particular setting. The one who reported that the patient was making excellent progress came from a setting (OT) in which the patient was indeed showing excellent progress, whereas the other came from a setting (ward) in which the patient had showed little progress. Luckily, our data, which came from monitoring the patient wherever he went, clarified the complex picture of place dependencies in the patient's status and progress and in staff judgments.

Long Time Periods

Many short-term mechanisms of good lead to harm in the long run. Sometimes the reverse is true. These reversals in the ecological domain occur because of the synergistic, cyclical workings of ecological systems over time. Traditional views of human behavior and its articulation with environmental systems seldom even consider the possibility of synergistic effects over long periods of time. It is unusual to ask whether our ways of putting persons into typical residential systems, penal and correctional institutions, schools, industrial facilities, and hospitals, or subjecting them to our favorite forms of treatment will lead to negative results that will be visited back upon them or their children over the long haul. Part of the problem here is that the short-term effects often appear pleasant, innocuous, or serene. The other part of the problem is that the very long time periods so important to ecological understanding seldom enter into our programs of research.

More generally, in keeping with the characteristics of behavior–environment systems and the kinds of behavioral dimensions with which the ecologist often works (e.g., adaptation, accomodation, functional achievement, long-range behavior, and sometimes even survival), the ecologist not only allows, but sometimes demands, unusually long periods and time dependencies in his research. For generations, behavioral scientists involved in development—child development, personality development, cultural development—have tried to demonstrate how events at one time can function as remote antecedents of something that occurs later (e.g., amount of social and environmental stimulation during infancy and childhood as a remote antecedent of later intelligence). Over and above this traditional search for early antecedents, the ecologist advocates truly longitudinal research that monitors interdependencies continuously, or nearly continuously, for extended periods of time. At least three concerns lead to this emphasis. First, sequential phenomena that emerge over time are among the most important properties of behavior (Barker, 1963b; Nesselroade and Reese, 1973; Raush, 1969). Second, ecosystems in general (and, therefore, probably behavioral–ecological systems) follow rules of succession and internal distribution in which the phases and changes in the systems over time become the critical issue. Third, we know by now from other areas (e.g., crop diseases, public health problems, pollution, insecticides) that empirical monitoring of very long sequences can be both scientifically illuminating and pragmatically critical. The behavioral ecologist would argue that human behavior and its emergent relations to its contexts must also be studied in such long-term perspective. The fact that the behavioral sciences cannot match the sophistication of ecological biochemistry or agronomy is no excuse to wait.

Of course, not all of the important phenomena of human functioning and behavior require very long periods of study. The *ecological* problem is that behavioral scientists are distressingly unwilling to differentiate those behavioral issues that require the emergent time frame from those that do not and even less willing to knuckle down and tackle those that do. Many psychologists would much rather do several discrete studies with many subjects each year than work on a phenomenon whose important properties unravel and cycle over a period of months or years. Research with very long time frames is absolutely necessary for the development of an ecological perspective on behavior.

Since 1973, we have been conducting our research on the aftermath of spinal cord injury in a longitudinal mode. The behavior of individual patients is monitored from admission to the hospital until discharge (an average of three months) and then beyond discharge for 12 months. At the time of this writing (fall, 1975), we have monitored complete hospital stays for seven patients. We plan to continue this work until we have finished monitoring such 15-month segments from the lives of 30 persons with spinal cord injuries.

We have undertaken this type of longitudinal study for several reasons. First, severe spinal cord injury is catastrophic and, if the person is going to survive and live a meaningful and functional life, then many complex changes must occur during the first year and a half after the injury. The only way to understand that complex process is to describe it as it occurs. Taken together, our data provide information on the long-term behavioral trajectory of spinal cord injured persons, well into the period of posthospital adjustment. Second, because they describe the gradual reemergence of behavioral adaptation to the environment following catastrophic loss, the data provide basic understanding of behavior–environment relations. Third, the data provide an ongoing set of behavioral indicators of patient progress and a new information base for staff members to use in making decisions about patient programs. Fourth, a treatment may appear successful in the short run and yet prove maladaptive in the long run. As yet, we do not know whether such striking inconsistencies will appear in the patient data. Should they occur, we will be in a good position to document them. Fifth, it is important that for the first time we can evaluate treatment programs in terms of long-range outcomes in patient behavior.

Figure 6 displays one of the characteristic longitudinal curves that can be extracted from out data. The figure shows the proportion of unaided performances on a weekly basis for three patients. Although the data cover complete hospital stays, keep in mind that for a patient being monitored information is available to the research staff or to hospital personnel at any given time during a patient's hospital stay.

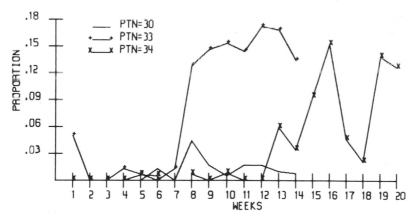

Figure 6. Proportion of unaided performances to total performances by three patients across their periods of hospitalization.

Figure 6 illustrates several important things. Although a definitive test must wait until data on more patients are available, our experience to date suggests several hypotheses. First, it appears that patients whose inhospital performance looks like patient 30 (little slope and low predischarge level) do less well after discharge than patients whose inhospital performance looks like patient 33 (early, rapid acceleration and high predischarge level). However, the pictures presented by patients 33 and 34 are even more interesting. By the time of discharge, patient 34 was performing at a level quite similar to patient 33. However, even with the similar predischarge level, it appears that patients whose inhospital performance looks like patient 34 (late onset of positive change and slower acceleration) do less well than patients like number 33 (early onset of change and rapid acceleration). Without continuous data gathered over a long period of time, such differences in performance and their relative predictive values would remain undiscovered.

Our longitudinal data also dispel some pet ideas and conventional wisdom. For example, it has often been assumed in rehabilitation that a decrease in idle time is a sensitive indicator of patient progress. In contrast, we find that proportion of idle time varies much less over time than do other indicators of patient performance (e.g., independence, active performance, behavioral diversity, environmental diversity).

One final procedural issue is pertinent here. If complex studies are to be maintained for long periods of time, then efficient techniques must be developed that enable long-term monitoring. We chose the method of narrative observation because we were working in an area with no precedents and in which little was known. However, the techniques we have used are expensive, burdensome, and time-consuming. Furthermore, if some derivative of our system for monitoring patient performance is to be developed for clinical use by hospital personnel, then the techniques for monitoring must be made more efficient and less cumbersome. We have taken two major steps in this direction. The first, a technique for obtaining the same data by more efficient means, involves refining the observational procedure by having observers work from a standard, time-cued form and make more judgments regarding behavior units on the spot (Crowley, 1976). With this technique, turnaround time from observation to usable data is almost immediate.

The second involves the development of instrumented monitoring. We have developed a system of pressure-sensitive pads placed under the patient's mattress and connected to a mechanism for continuous recording of movement on the bed surface and time out of bed. We have also developed simple mechanical odometers that are attached to the wheels of the patient's wheelchair. We find that time out of bed and wheelchair mobility appear to be excellent approximations or indices of some of our other behavior measures, e.g., number of settings entered, number of entries into settings, zest. The promising aspect here is that the instrumented measures (obtained easily and continuously) may serve as adequate substitutes for the direct but more complex measures of performance obtained with much more difficulty through observations.

Systems Concepts and the Illusion of Side Effects

It is common for the ecologist to couch much of what he does and thinks about in the terminology and concepts of systems. This happens for several reasons. Sometimes it is because of the extensiveness of the phenomena under study, sometimes because of their complexity or because of the emphasis on interdependence of many variables at many levels, sometimes because systems theory brings to bear an appropriate and powerful set of formal principles, and sometimes simply because *system* is the best metaphor or image the ecologist can conjure up to communicate what he is trying to say. As science and technology have grown in complexity, the use of systems theory and terminology has become more widespread. Thus, it is only reasonable that ecologists would turn to this discipline for tools of conceptual representation and analysis. There is a growing awareness that we need new and sophisticated forms of analysis and synthesis that are not amenable to the traditional, either-or methodology that permeates Western scientific thinking. Systems theory and its various derivatives offer the tools for representing *interdependence* and simultaneous, time-related complexity (Berrien, 1968; Bruhn, 1974; Buckley, 1968; Laszlo, 1972).

Ecosystem principles often can be dramatized best by demonstrating that a given intervention or change leads to complicated and unintended ramifications. The same is true for ecobehavioral systems. One of the significant features of many social, physical, and biological systems is that their components function as integrated wholes; manipulation of any part of such a system will affect each of the other parts and change the whole. Even the most positively motivated intrusions into interdependent systems can lead to all sorts of unanticipated effects, many of which are unpleasant and pernicious. To label these unintended effects as "side effects" only compounds the problem. Anticipated and unanticipated effects are all functions of the system's processes. Side effects do not occur in the real world. They exist in our images of events for which we *expect* some things and not others.

In our growing awareness of ecological phenomena, we are reluctant to introduce new biotic elements and new chemicals into our ecological systems, but we display dismal irresponsibility when it comes to intervening in behavioral and behavioral–environment systems. Almost every day we hear of projects or technologies being changed, slowed down, stopped, or disapproved on ecological grounds, because of the known complexity or delicacy of ecosystems. The location of a proposed factory is switched, a bridge is not built, a planned freeway is rerouted, a smokestack is modified, someone is restrained from introducing a new animal into an area, or a pesticide is taken off the market. Those who think that no changes are required in the behavioral sciences, or that arguments for a new effort in behavioral ecology are just so much window dressing should ask just one question: How often have I heard of a program or project whose target is human behavior being changed, slowed down, stopped, or disapproved on *behavioral*–ecological grounds, because of the *known* complexity and delicacy of ecobehavioral systems? The answer should sober us quickly.

It is quite foreign to psychologists (as well as many other social scientists) to think of the physical, social, and behavioral environment as inextricable parts of the behavioral processes of organisms and as relating to them in ways that are extremely complex. Two results that follow are (1) relatively splintered, bitlike attempts to understand large-scale, complex phenomena, and (2) blithe forms of bitlike tinkering in complex, interdependent systems. Every intervention, every human artifice, has its price, no matter how well-intentioned the agent of intervention may be. Unintended effects occurring in the short- or long-range outcome of a purposeful intervention into human behavior are viewed by the behavioral ecologist as important aspects of interdependent phenomena that need to be understood (Alexander *et al.*, 1976; Willems, 1973b, 1974).

Partly because of their rigor and partly because of the explicitness with which their intended effects can be spelled out, various programs of applied behavior analysis or behavior modification illustrate these problems very clearly for human behavior. Buell, Stoddard, Harris, and Baer (1968) used teacher attention to reinforce a withdrawn girl's use of play equipment in a preschool setting. The intervention was successful (intended effect), but it also affected the girl's interactions with other children, such as touching, verbalizations, and cooperative play (positive unintended effects). Wahler, Sperling, Thomas, Teeter, and Luper (1970) found that parents' successful efforts to reduce nonspeech deviant behaviors by their children also led to a reduction in stuttering and this positive unintended effect was not a function of differential reinforcement of stuttering and fluent speech. Twardosz and Sajwaj (1972) found that increasing the rate of sitting in a retarded and hyperactive boy also decreased the rate of weird posturings and increased the rate of proximity to other children. Drabman and Lahey (1974) used teacher feedback to reduce the rate of disruptive behavior by a child in a classroom, but they found that the rate of disruptive behavior by other children was reduced as well.

Unintended effects of a more unpleasant and disruptive sort also occur. Herbert, Pinkston, Hayden, Sajwaj, Pinkston, Cordua, and Jackson (1973) trained parents in the use of differential attention to increase appropriate behaviors and decrease deviant behaviors in six children. One child showed some improvement and one showed no change. In the cases of four children, dramatic, intense, and durable (but unintended) effects showed up, both in the treatment setting and other settings (e.g., assaulting mother, scratching self until bleeding, enuresis, throwing tables and chairs). The investigators were able to show that these effects occurred despite tenacious programs of differential reinforcement by parents.

Wahler (1975) found that the successful induction of planned treatment effects in one setting (home or school) was accompanied by unplanned effects in the other setting. Another clear example comes from the work of Sajwaj, Twardosz, and Burke (1972), who found various "side effects" of manipulating single behaviors in a preschool boy. They arranged for the teacher to ignore the child's initiated speech to her (nagging) in one setting of the preschool. This tactic was successful in reducing the nagging, but produced systematic changes in other behavior by the child in the same setting and in another setting as well. Some of the unintended effects were desirable (increasing speech initiated to children, increasing cooperative play) while some were undesirable (decreasing task-appropriate behavior, increasing disruptive behavior) and some were neutral (use of girls' toys). The investigators were able to show that the covarying effects were not due to differential attention by the teacher applied directly to those behaviors, but were somehow (as yet, mysteriously) a function of modifying another single dimension of behavior.

It seems clear that the simplistic models and technologies of change used by the applied behavior analyst are not comprehensive enough to lead to the understanding of complex human behavior or to predictable interventions. Extensive research should be conducted to ascertain which *kinds* of unintended effects occur most frequently in various settings and *why* they occur so that practitioners can begin to predict such effects and plan interventions with these effects in mind. It is embarrassing that behavioral ecologists cannot yet specify how such monitoring should be done. We do not know enough yet about person–behavior–environment systems, but finding out is vital to us all.

There are other examples that fall higher on a hierarchy of setting size and complexity. In a mental hospital ward, Proshansky *et al.* (chaps. 3 and 43, 1970b) used some amenities of interior design to increase the rate of sociable occupancy and use of a solarium and they found that the rate of detached, withdrawn, standing behavior went down. However, they had only succeeded in changing the *location* of the troubling behavior; a great deal of it now occurred at the other end of the corridor, by the nurses' station.

At the level of residential systems, the work of investigators such as Newman (1972) and Galle, Gove, and McPherson (1972), and the accumulation of experiences with arrangements such as Pruitt-Igoe in St. Louis all suggest that a torrent of troubling effects can follow from the concatenation of people in large groupings, high population density, and physical design whose behavioral principles are very poorly understood. More and more frequently, engineers, architects, and other agents of environmental planning and policy are turning to the social and behavioral sciences with various forms of the question: How should we design this environment to make it best suited for human inhabitants? We must confess that mostly we know after the fact that some things have *not* worked and we can offer almost no *prospective* prescriptions. For example, the new town of Skarholmen, near Stockholm in Sweden, was built to incorporate the most advanced architectural technology (Dubos, 1971). Among its 20,000 inhabitants, particularly among its children and teen-agers, there followed an alarming rate of restlessness, agressiveness, and withdrawal that was correlated with nearness to the center of town and the density of living arrangements. Behavioral scientists studying and evaluating the new town suggested that the antisocial behaviors were related to population density and the austere form of architecture. Whether this specific explanation is correct or not, it would appear that the unanticipated behavioral problems are symptomatic, not of personal problems, but of problems in the relationship of habitat and behavior.

While such examples offer strong reinforcement to the behavioral ecologist and some of his favorite arguments and beliefs, they are also troubling, for several reasons. First, they raise serious questions about some of the most promising, most explicit, and most powerful techniques of behavior change available today. Techniques of behavior modification usually are targeted and applied quite narrowly. In the typical case, they are applied to one subject, one or two settings, and few behaviors at a time and they are applied for a relatively brief period. The second issue is that if so many documentable unintended effects and ecological ramifications can occur with the use of such carefully targeted and short-term mechanisms of behavior change, then just imagine what must happen in the aftermath of much more broadly gauged, long-term, and diffuse programs and arrangements of behavior influence. Examples of the latter are: advancement and incentive programs in business and industry, community mental health programs, transportation systems, work schedules, educational arrangements, correctional and rehabilitation settings, residential systems, day care programs, and systems for delivery of health services. Finally, those of us who take an ecological perspective seriously have our work cut out for us. To date, we have functioned almost exclusively like Monday morning quarterbacks. Someone else tries something and we then swoop down crying, "I've gotcha with your unintended effects showing!" We are confronted with the monumental tasks of learning how to monitor and understand the network of influences in such systems and of developing anticipatory and prescriptive models and guidelines for intervention.

In our research program at TIRR, the system metaphor and the issue of labelling events as various kinds of effects have been centrally important. Between the 1968 and 1971 observations, the hospital's system for delivering care was changed. Through 1968 and until a few months before the observations in 1971, the treatment program and the resulting care delivery system were run along fairly traditional lines. The hospital's administration intruded into the system by establishing a Spinal Cord Center in 1971. The Center, which had its own location in the hospital, housed a group of 15 to 20 patients. A core team of persons from various staff categories was assigned to the Center and its patients, and the responsibilities and affiliations of the team resided with the Center patients rather than with their own professional groups and locations. Under this arrangement, all members of the Center's core team were to be directly involved in all aspects of a patient's program: diagnosis, assessment, treatment plan, delivery of treatment, and modification of treatment program. Underlying this arrangement was the assumption that the assignment of a multidiscipline core staff to a patient, to a location, and to a crossdisciplinary responsibility would enhance and optimize the care that patients received, with results in terms of individualization of care, better adaptation, better progress, shorter hospitalization time, better understanding of disability, and better preparation of the patient and family for life outside the hospital. It was assumed that there should follow a general intensification of the rehabilitation program and its independence training for patients and that through a more coordinated focusing of staff effort and involvement, patients should be caught up earlier in behavioral performances that were oriented toward their readjustment and functional independence. In other words, one of the purposes of the Spinal Cord Center was to *redistribute* patient behavior over the course of hospitalization.

The important point for present purposes is that our research operated under the guiding belief that a complex system involving persons, environment, facilities, and programs was being monitored and that the change to the Spinal Cord Center changed the relations among the components of that system. With a complicated set of data, the task was to document both anticipated and unanticipated changes and check them against the goals of the center.

First, some aspects of the distribution of patient behavior remained invariant through the change. Across all patients and across kinds of behavior, both the proportion of patient behavior that occurred in various parts or settings of the hospital and the frequencies of various kinds of behavior remained constant through the change to the Center. The number of treatment-related behavior settings stayed nearly constant. Furthermore, the proportion of time that patients spent in complete idleness remained the same through the change.

Second, a number of aspects of the system changed in positive or anticipated directions. Two examples illustrate these positive changes. From the observational data, it was possible to calculate the proportion of patient behaviors in which various kinds of other persons were directly, behaviorally involved (e.g., help with a transfer, help with eating a meal). Before the change to the center, aides and orderlies carried the largest burden of such involvements by far. One goal of the center was to change this distribution of involvements so that more advanced staff members would be directly involved with patients more often. Also, in the interests of better preparing patients and family members for life outside the hospital, another goal was to create conditions under which visitors, such as spouses and other

TABLE 4
PROPORTION OF PATIENT BEHAVIORS IN WHICH FOUR KINDS OF
OTHER PERSONS WERE DIRECTLY INVOLVED

	1968	1971	Percent change 1968 to 1971
Aides and orderlies	.41	.24	− 41%
Nurses	.11	.14	+ 27%
Physical therapists	.10	.14	+ 40%
Visitors	.04	.10	+150%

family members and friends, would become directly involved with patients more often. Table 4 indicates that these purposes were achieved to some extent. The rate of direct involvement with patients by aides and orderlies went down while the rates for nurses, physical therapists, and visitors went up.

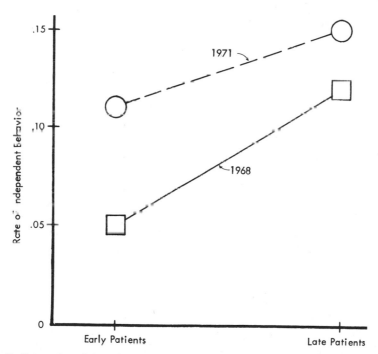

Figure 7. Proportion of time that early and late patients displayed behavioral independence in 1968 and 1971.

Another example of a positive change occurred in the independence of the patients (self-instigated and unaided behavior). Figure 7 shows the results before and after the introduction of the Center. As had been hoped, not only were patients generally more independent after the change, but they displayed the independence earlier.

Finally, against the background of constancies and some positive changes, the system also took back some things and produced some negative and unanticipated effects. These can be illustrated with two examples. One of the purposes of behavioral rehabilitation at the hospital is to prepare the patient for life outside the hospital. One way to do that is to make sure that as patients progress over time and regain some behavioral independence, they practice their new independence skills in more and more settings of the hospital and practice

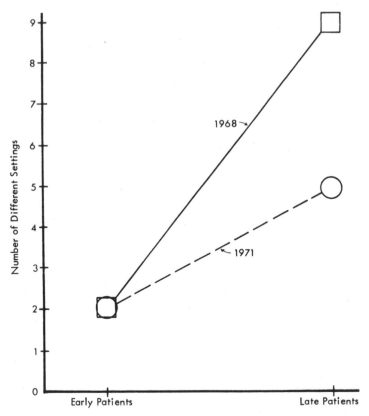

Figure 8. Number of different hospital settings in which early and late patients performed independently in 1968 and 1971.

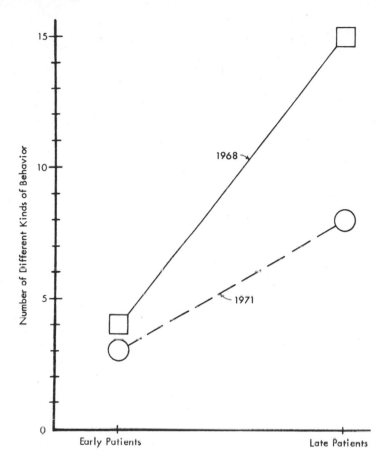

Figure 9. Number of different kinds of behavior that early and late patients performed
independently in 1968 and 1971.

carrying out many different behaviors in independent fashion. Figures
8 and 9 indicate that that was precisely what was happening in 1968,
before the change to the Center (see solid lines in both figures). In
1968, late patients performed independently (on their own initiative
and with no help from others) in far more of the hospital's settings
(Figure 8) and in a greater number of different kinds of behavior
(Figure 9) than early patients. However, this picture changed with the
creation of the Center; the environmental and behavioral range of the
late patients' independent performance became much more restricted
(dotted lines in both figures).

These data represent only a few discrete illustrations of a more complex analysis of a system made up of persons, facilities, programs, behaviors, and settings. However, the data illustrate how research that is guided by ecological principles can lead one to look for phenomena that might be overlooked otherwise. An intervention in the ongoing system resulted in some constancies and some gains. But these constancies and gains were also linked interdependently with losses in some other functions, losses that may not have been apparent unless the systemlike functioning of the hospital was assumed and unless extensive behavioral monitoring was carried out.

CONCLUDING COMMENTS

By means of some general principles, analogies, arguments, and illustrations, I have tried to communicate some of the essential flavor and current vision of behavioral ecology. Included in this vision are key ideas such as complexity and interdependency of man–environment relations across many levels, behavior–organism–environment systems, principles of reciprocity and exchange, cycles and emergent phenomena over long periods of time, and complicated ramifications of interventions and changes. I have argued that this vision has important methodological implications that fall into several major clusters of needed research: (1) new forms of complex monitoring and description of social, institutional, behavioral, and environmental systems; (2) extension of such monitoring over long periods of time; (3) extensive ecological monitoring of the aftermath of targeted interventions; (4) more careful monitoring of the effects of other, naturally occurring changes; (5) more careful documentation of the inextricable involvement of the environment in human performance; and (6) more explicit documentation of environmental dependencies in behavior.

Because of space limitations, some other important aspects of behavioral ecology have been left out (see Willems, 1973a, 1973c, 1976b; Alexander et al., 1976): (1) taxonomy and the problems of classifying environments and behaviors; (2) transdisciplinary and multilevel research and theory; (3) emphasis on habitability and the right .of human beings to good environments; (4) methodological issues of evaluative research; and (5) the importance of ecological diagnosis (i.e., system-based versus person-based diagnosis).

Next, by means of a few discrete examples from an extended, ongoing program of research, I have tried to illustrate some of the ways in which the ecological vision can guide and affect research on a real-life problem of human performance. At the beginning of this decade, Odum (1971) pointed out that ". . .just how applied human ecology is to be developed and structured so that worthwhile goals can be achieved in the real world of society can be but dimly perceived at this point of history" (p. 510). I hope I have illustrated that Odum's skepticism is no longer necessary. For both the formulation of the ecobehavioral perspective and for the study of the microecology of human performance, we now have much more than "dim perceptions." We have some articulated principles and guidelines and we have some paradigms for research on persons in institutional settings where environmental aspects of performance are central issues. Currently, the perspective and methods of our research program are being applied by other investigators in studies of mental retardation, hyperactivity, ambulatory care facilities, medicine wards in a general hospital, and spinal injury sections of Veteran's Administration hospitals.

By now, the perspective of behavioral ecology has developed to the point that we can lay to rest several classic issues of human science. First, the ecologist must strive to play down the distinction between "basic" and "applied" research that has been such an albatross around the neck of behavioral scientists. The agronomist needs *good* research on the ecology of plant life; he does not need the burden of worrying about whether a study of the cytogenetics of corn is "basic" while a study of wheat rust is "applied." Likewise, behavioral ecologists need *good* research on the ecology of human behavior; they do not need the burden of classifying a study of the concept of privacy as "basic" and a study of the effects of school size on children as "applied."

Second, we often hear the protest that science is supposed to be the search for *simplifying* principles. Simon (1969) notes, "The central task of a natural science is to make the wonderful commonplace: to show that complexity, correctly viewed, is only a mask for simplicity" (p. 1). That may be so in the very long view. However, if the arguments of behavioral ecology are tenable, then the masking picture must be applied with care and with impeccable timing. In the work of behavioral scientists who deal with behavior–organism–environment relations, the masking process usually is more like the act of putting on blinders; the *assumption* that the phenomena are simple is made before the search for simplicity even begins. It just may be that, in the long run, the most direct and efficient path toward scientific understanding of behavior will involve the timely recognition and acceptance of complexity within an ecological perspective.

There probably will be various forms of resistance by behavioral scientists to perspectives like behavioral ecology that, if they are accepted, should tincture our consciousness, our views of human behavior, and our work. The story is told of a peasant who lived in an area for which the political boundaries were unclear; it might be in Poland or it might be in Russia (Mitchell, 1967). The peasant was deeply concerned about which it was going to be and so, when a surveying party came through and ran their lines, he was out there with them, watching every move. A member of the surveying party asked the peasant why he was so concerned and he said it was because he wanted badly to know whether he would be living in Poland or Russia. They asked him why he was so concerned about that; whether his concern might be over losing some personal or property rights. The peasant said, "No, no. I don't care about all those things. I just never could stand those awful Russian winters."

Confronted with the professional challenge and the opportunity to deal with the ecological complexity of human behavior and human functioning, many persons respond in analogous fashion. They yearn for some authority who will declare that simple models and simple strategies of research will do the job so that they can breathe a sigh of relief—"I can't stand all that awful complexity, anyway." However, declarations and labels do not change either the phenomena of human behavior or the emerging demand for growth and expansion of perspective. Stated more positively, the ecological perspective on behavior represents a very exciting opportunity, an opportunity to generate altogether new questions, to formulate completely new modes of scientific work, and, most importantly, to develop nontrivial theories of behavior.

ACKNOWLEDGMENTS

Work on this paper and on the research reported was supported by Research and Training Center No. 4 (RT-4), Texas Institute for Rehabilitation and Research and Baylor College of Medicine, funded by the Rehabilitation Services Administration, USDHEW.

REFERENCES

Alexander, J. L. The wheelchair odometer as a continuous, unobstrusive measure of human behavior. Unpublished masters thesis, University of Houston, 1976
Alexander, J. L., Dreher, G. F., and Willems, E. P. Behavioral ecology and humanistic and behavioristic approaches to change. In A. Wandersman, P. Poppen, and D. Ricks (Eds.), *Human behavior and change*. Elmsford, N.Y.: Pergamon, 1976, in press.

Altman, I. *The environment and social behavior.* Monterey, Calif.: Brooks/Cole, 1975.

Bakeman, R., and Helmreich, R. Cohesiveness and performance: Covariation and causality in an undersea environment. *Journal of Experimental Social Psychology,* 1975, *11,* 478–489.

Baker, C. N. Surveys of environmental negotiability in homes of disabled persons. Unpublished masters thesis, University of Houston, 1976.

Barker, R. G. On the nature of the environment. *Journal of Social Issues,* 1963a, *19,* No. 4, 17–38.

Barker, R. G. (Ed.) *The stream of behavior.* New York: Appleton-Century-Crofts, 1963b.

Barker, R. G. Explorations in ecological psychology. *American Psychologist,* 1965, *20,* 1–14.

Barker, R. G. *Ecological psychology.* Stanford, Calif.: Stanford University Press, 1968.

Barker, R. G. Wanted: An eco-behavioral science. In E. P. Willems and H. L. Raush (Eds.), *Naturalistic viewpoints in psychological research.* New York: Holt, Rinehart and Winston, 1969, 31–43.

Barker, R. G., and Gump, P. V. (Eds.) *Big school, small school.* Stanford, Calif.: Stanford University Press, 1964.

Barker, R. G., and Schoggen, P. *Qualities of community life.* San Francisco: Jossey-Bass, 1973.

Barker, R. G., and Wright, H. F. *Midwest and its children.* New York: Harper & Row, 1955.

Benarde, M. A. *Our precarious habitat.* New York: W. W. Norton, 1970.

Berrien, F. K. *General and social systems.* New Brunswick, N.J.: Rutgers University Press, 1968.

Bijou, S. W., Peterson, R. R., and Ault, M. H. A method to integrate descriptive and experimental field studies at the level of empirical concepts. *Journal of Applied Behavior Analysis,* 1968, *1,* 175–191.

Brandt, R. M. *Studying behavior in natural settings.* New York: Holt, Rinehart and Winston, 1972.

Breland, K., and Breland, M. *Animal behavior.* New York: Macmillan, 1966.

Bruhn, J. G. Human ecology: A unifying science? *Human Ecology,* 1974, *2,* 105–125.

Buckley, W. (Ed.) *Modern systems research for the behavioral scientist.* Chicago: Aldine, 1968.

Buell, J., Stoddard, P., Harris, F. R., and Baer, D. M. Collateral social development accompanying reinforcement of outdoor play in a preschool child. *Journal of Applied Behavior Analysis,* 1968, *1,* 167–173.

Chapanis, A. The relevance of laboratory studies to practical situations. *Ergonomics,* 1967, *10,* 557–577.

Chase, A. *The biological imperatives.* New York: Holt, Rinehart and Winston, 1971.

Colinvaux, P. A. *Introduction to ecology.* New York: Wiley, 1973.

Cowen, E. L. Social and community interventions. *Annual Review of Psychology,* 1973, *24,* 423–472.

Craik, K. H. Environmental psychology, In *New directions in psychology—IV.* New York: Holt, Rinehart and Winston, 1970, 1–121.

Crowley, L. R. Development and assessment of an alternative to the narrative observation. Unpublished masters thesis, University of Houston, 1976.

Danford, G. S., and Willems, E. P. Subjective responses to architectural displays: A question of validity. *Environment and Behavior,* 1975, *7,* 486–516.

Drabman, R. S., and Lahey, B. B. Feedback in classroom behavior modification: Effects on the target and her classmates. *Journal of Applied Behavior Analysis,* 1974, *7,* 591–598.

Dreher, G. F. Reliability assessment in narrative observations of human behavior. Unpublished masters thesis, University of Houston, 1975.

Dubos, R. *Man adapting*. New Haven, Conn.: Yale University Press, 1965.

Dubos, R. *So human an animal*. New York: Charles Scribner's Sons, 1968.

Dubos, R. The despairing optimist. *American Scholar*, 1971, 40, No. 4 (Autumn), 565–572.

Ebbesen, E. B., and Haney, M. Flirting with death: Variables affecting risk taking at intersections. *Journal of Applied Social Psychology*, 1973, 3, 303–324.

Elms, A. C. The crisis of confidence in social psychology. *American Psychologist*, 1975, 30, 967–976.

Fawl, C. L. Disturbances experienced by children in their natural habitats. In R. G. Barker (Ed.), *The stream of behavior*, New York: Appleton-Century-Crofts, 1963, 99–126.

Galle, O. R., Gove, W. R., and McPherson, J. M. Population density and pathology: What are the relations for man? *Science*, 1972, 176, 23–30.

Gump, P. V. Intra-setting analysis: The third grade classroom as a special but illustrative case. In E. P. Willems and H. L. Raush (Eds.), *Naturalistic viewpoints in psychological research*. New York: Holt, Rinehart and Winston, 1969, 200–220.

Gump, P. V., and Kounin, J. S. Issues raised by ecological and "classical" research efforts. *Merrill-Palmer Quarterly*, 1959–1960, 6, 145–152.

Herbert, E. W., Pinkston, E. M., Hayden, M. L., Sajwaj, T. E., Pinkston, S., Cordua, G., and Jackson, C. Adverse effects of differential parental attention. *Journal of Applied Behavior Analysis*, 1973, 6, 15–30.

Insel, P. M., and Moos, R. H. (Eds.) *Health and the social environment*. Lexington, Mass.: Lexington Books, 1974.

Ittelson, W. H., Proshansky, H. M., Rivlin, L. G., and Winkel, G. H. *An introduction to environmental psychology*. New York: Holt, Rinehart and Winston, 1974.

Ittelson, W. H., Rivlin, L. G., and Proshansky, H. M. The use of behavioral maps in environmental psychology. In H. M. Proshansky, W. H. Ittelson, and L. G. Rivlin (Eds.), *Environmental psychology*. New York: Holt, Rinehart and Winston, 1970, 658–668.

Johnson, S. M., and Bolstad, O. D. Methodological issues in naturalistic observation: Some problems and solutions for field research. In L. A. Hammerlynck, L. C. Handy, and E. J. Mash (Eds.), *Behavior change: Methodology, concepts and practice*. Champaign, Ill.: Research Press, 1973, 7–67.

King, J. A. Ecological psychology: An approach to motivation. In W. J. Arnold and M. M. Page (Eds.), *Nebraska symposium on motivation*. Lincoln, Neb.: University of Nebraska Press, 1970, 1–33.

Laszlo, E. *The systems view of the world*. New York: Braziller, 1972.

LeCompte, W. F. The taxonomy of a treatment environment. *Archives of Physical Medicine and Rehabilitation*, 1972, 53, 109–114.

LeCompte, W. F., and Willems, E. P. Ecological analysis of a hospital: Location dependencies in the behavior of staff and patients. In J. Archea and C. Eastman (Eds.), *EDRA-2: Proceedings of the 2nd annual environmental design research association conference*. Pittsburgh: Carnegie-Mellon University, 1970, 236–245.

Margalef, R. *Perspectives in ecological theory*. Chicago: University of Chicago Press, 1968.

McGuire, W. J. Theory-oriented research in natural settings: The best of both worlds for social psychology. In M. Sherif and C. W. Sherif (Eds.), *Interdisciplinary relationships in the social sciences*. Chicago: Aldine, 1969, 21–51.

McHale, J. *The ecological context*. New York: Braziller, 1970.

Menzel, E. W., Jr. Naturalistic and experimental approaches to primate behavior. In E. P. Willems and H. L. Raush (Eds.), *Naturalistic viewpoints in psychological research.* New York: Holt, Rinehart and Winston, 1969, 78–121.

Miller, R. H., and Keith, R. A. Behavioral mapping in a rehabilitation hospital. *Rehabilitation Psychology*, 1973, *20*, 148–155.

Mischel, W. *Personality and assessment.* New York: Wiley, 1968.

Mitchell, R. B. It depends: Or how does the physical city affect people? In V. C. Vaughan, III (Ed.), *Issues in human development.* Washington, D.C.: National Institute of Child Health and Human Development, 1967, 164–179.

Moos, R. H. Conceptualizations of human environments. *American Psychologist*, 1973, *28*, 652–665.

Moos, R. H. Systems for the classification of human environments: An overview. In R. H. Moos and P. M. Insel (Eds.), *Issues in human ecology.* Palo Alto, Calif.: National Press Books, 1974, 5–28.

Moss, G. E. *Illness, immunity, and social interaction.* New York: Wiley, 1973.

Nesselroade, J. R., and Reese, H. W. (Eds.) *Life-span developmental psychology: Methodological issues.* New York: Academic Press, 1973.

Newman, O. *Defensible space.* New York: Macmillan, 1972.

Odum, E. P. *Fundamentals of ecology.* 3rd ed. Philadelphia: Saunders, 1971.

Pierce, L. M. A patient-care model. *American Journal of Nursing*, 1969, *69*, 1700–1704.

Powers, W. T. Feedback: Beyond behaviorism. *Science*, 1973, *179*, 351–356.

Proshansky, H. M., Ittelson, W. H., and Rivlin, L. G. The influence of the physical environment on behavior: Some basic assumptions. In H. M. Proshansky, W. H. Ittelson, and L. G. Rivlin (Eds.), *Environmental psychology.* New York: Holt, Rinehart and Winston, 1970a; 27–37.

Proshansky, H. M., Ittelson, W. H., and Rivlin, L. G. (Eds.), *Environmental psychology.* New York: Holt, Rinehart and Winston, 1970b.

Raush, H. L. Naturalistic method and the clinical approach. In E. P. Willems and H. L. Raush (Eds.), *Naturalistic viewpoints in psychological research.* New York: Holt, Rinehart and Winston, 1969, 122–146.

Raush, H. L., Dittmann, A. T., and Taylor, T. J. The interpersonal behavior of children in residential treatment *Journal of Abnormal and Social Psychology*, 1959a, *50*, 9–26.

Raush, H. L., Dittmann, A. T., and Taylor, T. J. Person, setting and change in social interaction. *Human Relations*, 1959b, *12*, 361–379.

Raush, H. L., Farbman, I., and Llewellyn, L. G. Person, setting and change in social interaction: II. A normal-control study. *Human Relations*, 1960, *13*, 305–333.

Rhodes, W. C. An overview: Toward synthesis of models of disturbance. In W. C. Rhodes and M. L. Tracy (Eds.), *A study of child variance.* Ann Arbor, Mich.: University of Michigan Press, 1972, 541–600.

Ronnebeck, R. W. A naturalistic investigation of community adjustment of facially disfigured burned teen-agers. Unpublished doctoral dissertation, University of Houston, 1972.

Sajwaj, T., Twardosz, S., and Burke, M. Side effects of extinction procedures in a remedial preschool. *Journal of Applied Behavior Analysis*, 1972, *5*, 163–175.

Schmitt, R. C. Some ecological variables of community adjustment in a group of facially disfigured burned children. Unpublished doctoral dissertation, University of Houston, 1971.

Sells, S. B. Ecology and the science of psychology. In E. P. Willems and H. L. Raush (Eds.), *Naturalistic viewpoints in psychological research.* New York: Holt, Rinehart and Winston, 1969, 15–30.

Shepard, P., and McKinley, D. (Eds.) *The subversive science: Essays toward an ecology of man.* Boston: Houghton Mifflin, 1969.

Simon, H. A. *The sciences of the artificial.* Cambridge, Mass.: M.I.T. Press, 1969.

Smith, R. L. *Ecology and field biology.* New York: Harper and Row, 1966.

Stuart, D. G. A naturalistic study of the daily activities of disabled and nondisabled college students. Unpublished masters thesis, University of Houston, 1973.

Thoresen, R. W., Krauskopf, C. J., McAleer, C. A., and Wenger, H. D. The future for applied psychology: Are we building a buggy whip factory? *American Psychologist,* 1972, *27,* 134–139.

Turnbull, C. M. *The mountain people.* New York: Simon and Schuster, 1972.

Twardosz, S., and Sajwaj, T. Multiple effects of a procedure to increase sitting in a hyperactive, retarded boy. *Journal of Applied Behavior Analysis,* 1972, *5,* 73–78.

Wahler, R. G. Some structural aspects of deviant child behavior. *Journal of Applied Behavior Analysis,* 1975, *8,* 27–42.

Wahler, R. G., Sperling, K. A., Thomas, M. R., Teeter, N. C., and Luper, H. L. The modifaction of childhood stuttering: Some response-response relationships. *Journal of Experimental Child Psychology,* 1970, *9,* 411–428.

Wallace, B. *Essays in social biology (Three volumes).* Englewood Cliffs, N.J.: Prentice-Hall, 1972.

Watt, K. E. F. (Ed.) *Systems analysis in ecology.* New York: Academic Press, 1966.

Watt, K. E. F. *Ecology and resource management.* New York: McGraw-Hill, 1968.

Watts, A. The individual as man-world. In P. Shepard and D. McKinley (Eds.), *The subversive science: Essays toward an ecology of man.* Boston: Houghton Mifflin, 1969, 139–148.

Wicker, A. W. Attitudes versus actions: The relationship of verbal and overt behavioral responses to attitude objects. *Journal of Social Issues,* 1969, *25,* No. 4, 41–78.

Wicker, A. W. An examination of the "other variables" explanation of attitude–behavior inconsistency. *Journal of Personality and Social Psychology,* 1971, *19,* 18–30.

Wicker, A. W. Processes which mediate behavior-environment congruence. *Behavioral Science,* 1972, *17,* 265–277.

Widmer, M. L. Telephone contact as a method for gathering data on the everyday behavior of noninstitutionalized adults. Unpublished masters thesis, University of Houston, 1976.

Willems, E. P. An ecological orientation in psychology. *Merrill-Palmer Quarterly,* 1965, *11,* 317–343.

Willems, E. P. Behavioral validity of a test for measuring social anxiety. *Psychological Reports,* 1967, *21,* 433–442.

Willems, E. P. Planning a rationale for naturalistic research. In E. P. Willems and H. L. Raush (Eds.), *Naturalistic viewpoints in psychological research.* New York: Holt, Rinehart and Winston, 1969, 44–71.

Willems, E. P. The interface of the hospital environment and patient behavior. *Archives of Physical Medicine and Rehabilitation,* 1972a, *53,* 115–122.

Willems, E. P. Place and motivation: Complexity and independence in patient behavior. In W. J. Mitchell (Ed.), *Environmental design: Research and practice.* Los Angeles: University of California at Los Angeles, 1972b, 4-3-1 to 4-3-8.

Willems, E. P. Behavioral ecology and experimental analysis: Courtship is not enough. In J. R. Nesselroade and H. W. Reese (Eds.), *Life-span developmental psychology: Methodological issues.* New York: Academic Press, 1973a, 195–217.

Willems, E. P. Go ye into all the world and modify behavior: An ecologist's view. *Representative Research in Social Psychology,* 1973b, *4,* 93–105.

Willems, E. P. Behavior–environment systems: An ecological approach. *Man–Environ-ment Systems,* 1973c, *3,* No. 2, 79–110.

Willems, E. P. Behavioral ecology as a perspective for man–environment research. In W. F. E. Preiser (Ed.), *Environmental design research.* Vol. 2. Stroudsburg, Pa.: Dowden, Hutchinson and Ross, 1973d, 152–165.

Willems, E. P. Behavioral technology and behavioral ecology, *Journal of Applied Behavior Analysis,* 1974, *7,* 151–165.

Willems, E. P. Relations of models to methods in behavioral ecology. In H. McGurk (Ed.), *Ecological factors in human development.* New York: Academic Press, 1976a, in press.

Willems, E. P. Behavioral ecology as a perspective for research in psychology. In C. W. Deckner (ed.), *Methodological perspectives for behavioral research.* Springfield, Ill.: Charles C. Thomas, 1976b, in press.

Willems, E. P., and Campbell, D. E. Behavioral ecology: A new approach to health status and health care. In B. Honikman (Ed.), *Responding to social change.* Strouds-burg, Pa.: Dowden, Hutchinson and Ross, 1975, 200–210.

Willems, E. P., and Crowley, L. R. *Narrative recording, structural coding, and content analysis of human behavior.* Houston, Texas: Texas Institute for Rehabilitation and Research, 1976.

Willems, E. P., and Vineberg, S. E. *Procedural supports for the direct observation of behavior in natural settings.* Houston, Texas: Texas Institute for Rehabilitation and Research, 1970.

Wright, H. F. *Recording and analyzing child behavior.* New York: Harper & Row, 1967.

Wright, H. F. *Children's behavior in communities differing in size.* 5 vols. Lawrence, Ks.: University of Kansas, 1969–1970.

7

Environmental Change and the Elderly

KERMIT K. SCHOOLER

Projections of future events may have become more frequent simply because a new century is approaching. Or, quite possibly, the incidence of such projections has increased because the technology for making them has proliferated and improved. Whatever the reason may be, one such set of projections was recently undertaken and presented by a group of scholars interested in the problems of aging (Neugarten, 1975). As background for the more sophisticated projections, one estimate predicted that in the year 2000 over 35 million people in the United States would be 65 years of age or over, in contrast to about 22 million at the time of this writing. The magnitude of this increase in the United States in the aged population is, to say the least, impressive. On a worldwide basis, the figures are even more impressive. And yet, as Beattie has pointed out, a number of significant futuristic studies have for all intents and purposes ignored or have omitted consideration of the elderly segment of the population (Beattie, 1975).

It should be made clear at the outset that this chapter is not principally a futures study. We may be sure that at the descriptive level, various indicators will change significantly over time: the age distribution, the man-made environment, the life-style. We should have confidence, however, that while the distribution in such classes

KERMIT K. SCHOOLER · School of Social Work, Syracuse University, Syracuse, New York.

of variables will change over time, the nature of relationships among them ought to be considerably more stable. This chapter, then, will address itself to the impact of changing environment on older people. The chapter will have the following tasks: The first will be to describe the elderly and their environment. The second major task will be to discuss what we know about the relations between environmental change and aging on the basis of a review of empirical findings. Finally, this chapter will address itself to theory, with the firm belief that good theory will serve as the guide for future research, perhaps enhancing not only our understanding of these natural phenomena but also our ability to predict and therefore control the future.

THE ELDERLY—A POPULATION AT RISK

A dean in a large university, discussing a prospective employee with one of his colleagues recently asked, "How old is he?" "What difference does it make?" the colleague replied. What difference, indeed? There was in this instance no reason to believe that the candidate's ability to perform was in any manner related to the candidate's chronological age. But the glib manner in which the inquiry was made and the startled response occasioned by the reply served to remind the dean one more time about the age-graded nature of our society. Moreover, in this instance it served the purpose of emphasizing the stereotypic nature of our beliefs about all age categories and about the elderly in particular. This is not to say that all of the beliefs convey only a positive or only a negative affect. As is the case with many systems of stereotype, some of the elements are contradictory to each other. The point to be made is simply this: Our attitudes, behaviors, and expectations toward the elderly stem from this system of stereotypic beliefs and, according to some writers, that system has a mythlike quality. Manney reminds us how our language betrays some of the underlying negative, if not outright hostile, attitudes toward aging and the elderly through such expressions as "old coot" and "old fool" or through aphorisms such as "you can't teach an old dog new tricks," while at the same time glossing over and obscuring these feelings through such labels as "the golden years" and "senior citizen" (Manney, 1975). Butler provides us with a fairly elaborate stereotypic sketch of old age:

> An older person thinks and moves slowly. He does not think as he used to or as creatively. He is bound to himself and to his past and can no longer change or grow. He can learn neither well nor swiftly, and, even if he

could, he would not wish to. Tied to his personal traditions and growing conservatism, he dislikes innovations and is not disposed to new ideas. Not only can he not move forward, he often moves backward. He enters a second childhood, caught up in increasing egocentricity and demanding more from his environment than he is willing to give it. Sometimes he becomes an intensification of himself, a character of a lifelong personality. He becomes irritable and cantankerous, yet shallow and feeble. He lives in his past. He is behind the times. He is aimless and wandering of mind, reminiscing and gregarious. Indeed he is a study in decline, the picture of mental and physical failure. He has lost and cannot replace friends, spouse, job, status, power, influence, income. He is often stricken by diseases which in turn restrict his movement, his enjoyment of food, the pleasures of well-being. He has lost his desire and capacity for sex. His body shrinks, and so too does the flow of blood to his brain. His mind does not utilize oxygen and sugar at the same rate as formerly. Feeble, uninteresting, he awaits his death, a burden to society, to his family and to himself. (Butler, 1975)

While Butler's stereotyped description stresses the negative, it should be pointed out that some stereotypes convey a positive affect as well. Elderly women, for example, are "grandmotherly" and therefore "kindly." In addition, we may attribute wisdom commensurate with the years, accumulated skills, and a storehouse of information. Of course, regardless of the affect associated with our stereotypes, when the observable reality runs contrary to the view held, some process must take place to justify the apparent contradiction between myth and reality. The "swinging" grandmother may become "cute" on the one hand or "disgraceful" on the other. But it should be clear to the reader by now that the problem with stereotypes about the elderly is the problem with all stereotypes and such generalizations. They are correct in some instances, but fundamentally it is the tremendous variety in the object class that gives lie to the stereotype. Indeed, there are those who will assert that the variation among the elderly exceeds the variation among younger age groups. And why should this not be so? For, in addition to all the variations genetically produced, there is the added variation due to the wealth of life's experiences, the hopes, the fears, the achievements, the disappointments, met and unmet, realized and unrealized throughout a lifetime. As this chapter moves into a cursory discussion of what we know of the psychological and biological makeup of the older person, the essential variability of the older person should be borne in mind. An understanding of that lack of uniformity will allow us to place in proper perspective later the main discussion of the relation between environmental change and aging.

The literature on psychological and physiological change with age confirms the view that the process of aging varies among individuals. Maddox and Douglas (1974) support this contention, stating, "development, change, and growth continue through the later years of the life span in spite of the decrement of social psychological, and physiological functioning which typically accompanies the aging process."

Psychological Change

Functions decline at varying rates, and performance capacities diminish with old age, as the general rule. For example, vision is dimmed with time, a process of change that begins even before middle age: A 50-year-old person needs twice as much light to see as well as a person of 20. Hearing, another sense that frequently declines long before old age, is usually only impaired in the high frequency range in old age. But this in itself is enough for the old person to miss numerous environmental cues, such as doorbells, various speech pitches, etc. The combination of the hearing and vision deficit fosters a sense of isolation, a reaction to anxiety and fear that are created when environmental cues are absent or not as acute. Although these are the major sensory deficits identified in the elderly, other senses such as taste, balance, and sensitivity to touch also diminish.

As with a general dulling of the senses, the elderly also experience a general slowing of response to sensory stimuli. This process is not well understood, except that the processing mechanisms of the brain are slowed: Simply, reception, integration, and execution in response to the stimuli take longer for elderly than for younger persons. Although such slowing occurs, studies indicate a difference between speed of response on the one hand, and accuracy on the other. Apparently an older person who has adequate time can perform a task just as well as a younger person. The older person seems to have more of a need to check responses and is less satisfied with rough accuracy. According to Manney, "Except when there is organic brain damage. . .mental processes work well for normal situations. What we lose with age is speed." (Manney, 1975).

As with the senses and speed of response, learning and intelligence have stereotypically been thought to decline in the elderly. Past research has supported this belief. Recently, however, the evidence indicates that intellectual performance among the elderly is hinged on social, health, and educational factors. Educational confounding is demonstrated by the similarity of findings when comparing the learning curve with the educational level curve for the elderly. Terminal drop, a phenomenon discussed by Riegel, clearly shows the confounding of intellectual performance by health (Riegel, 1972). The logical conclusion is that if health is maintained, so is intellectual ability.

Baltes and Schaie (1974) have done a most thorough investigation of intelligence in later life. Of the four measures of intelligence studied, they found that only one—visuo-motor flexibility—declined. Two measures, cognitive flexibility and visualization, did not decline at all, while the fourth measure, crystallized intelligence, actually improved.

Botwinick (1973), Canestrari (1963), and Eisdorfer (1965) suggest that performance deficits are a result of noncognitive as well as cognitive factors. For example, some performance deficits occur because of decline in response speed, but older adults exhibit a smaller deficit if they are given sufficient exposure to the stimulus. Other noncognitive factors to be considered when assessing performance are overarousal and meaningfulness.

Memory, as intelligence, does not decline with age in all areas. Smith (1975) found an interaction between age and memory that involves differential recall from long-term memory. Short-term recall was relatively unaffected by age. Other researchers have explained memory using a retrieval explanation. It is not the storage capacity that declines with age, but as Anders, Fozard, and Lillyquist (1972) show, it is the rate at which the information can be retrieved from memory that declines.

PHYSIOLOGICAL CHANGE

It has not been clearly shown whether biological changes are caused by environmental stresses, nor whether biological changes are independent of physiological changes or they exacerbate one another. Behavior has been interpreted in terms of the biological decrement model; aging is then represented by the decline of all physiological systems (Woodruff, 1973). Devries, finding that moderate exercise improves physiological functioning in the aged, suggests that it is not age itself that causes individuals to decline physiologically (De Vries, 1975).

Other research indicates that changes in physiological functions seemingly related to age may not be fixed and inevitable. Possibly behavioral and biological intervention may reverse the performance observed in old people. Although some decrements are identified with age, the pace of individual aging depends on genetics, nutrition, activity, and the psychosocial environment of the individual.

While some research demonstrates a physiological decline with age, it is invalid to consider these changes totally irreversible. For example, neuromuscular performance decreases as a function of increasing age, but simple reaction time and discrimination reaction time can both increase with age. Spirduso (1975) explains this potential increase as a result of physical activity. Activity may play a more dominant role in determining reaction and movement time speed than age.

Social Deficit

As if it were not sufficient that there are the decrements just noted (occurring, albeit, at varying rates), there is yet another class of "insults" to be sustained by the elderly: In addition to the negative stereotypes frequently held, older people in the United States make the transition from middle age with few, if any, of the positive societal supports given to other age changes or to other role transitions. As Rosow informs us in his cogent analysis of the process of socialization to old age (Rosow, 1974), not only are the elderly victimized by the negative stereotypes referred to earlier, but they are treated with indifference or rejection as members of a devalued class. Moreover, society expects a withdrawal from family, work, and other roles that at earlier age gave meaning and a sense of worth to the adult's life. A particularly poignant comment by Rosow contrasts the plight of the elderly with that of members of other devalued—or outcast—classes in our society, observing that the latter may know "redemption"—the criminal pays a debt to society, for example—but for the elderly there is no reversal or redemption.

In summary, because of psychological, physiological, and social losses, older people, as a class, are vulnerable to a multitude of threatening, and potentially injurious circumstances in their surrounding world.

Environmental Change

It has been noted that at various levels of government encouragement is given to those activities that will change the environment likely to impact the elderly: Housing is constructed, presumably especially for their benefit; old neighborhoods are demolished; services for the elderly are provided, often in a physical setting especially designed; roads are built, others aren't; the familiar town or city changes. Trite as it seems, one should periodically be reminded that when new structures (houses, hospitals, etc.) are built, it is an almost sure bet that someone will, voluntarily or involuntarily, move into them; or someone, voluntarily or involuntarily, will be enveloped by the changed environment: new housing, deteriorated housing, highways, parks, whatever. It seems appropriate in the light of (1) the increase in numbers of elderly, (2) the vulnerability of the elderly, and (3) the magnitude of environmental change impinging on the elderly that attention be directed toward understanding the impact of environmental change on the elderly.

Consider, first, the nature of the relation between the characteristics of the man-made environment on the one hand, and aspects of the aging process on the other. There are several current points of view that determine how some people will conceive of that relationship. Some, for example, will see the relations in terms of causation. In some sense, according to this view, elements or attributes of the environment are seen as "causing" aspects of the aging process. Hypothetically, according to this view, quality of housing may have an effect on physical health. A second kind of view recognizes that man affects his environment and in turn is affected by it. According to this view, the process may be seen as unending, and rather than being unidirectional (from "environment" to "behavior"), a feedback loop exists. At the *macro* level, according to this view, the natural environment, let us say, yields up mineral substances that are then utilized by people in such a way that the by-products "pollute" the natural environment, thereby having an impact on the physical health of human beings, who in turn will alter the physical environment so as to dispose of the pollutants, and so on. At the *micro* level, one might note that a person moving into a furnished apartment may alter the design of the interior in a manner quite contrary to the designer's intention and thereby enhance or inhibit behaviors that might be beneficial to the resident. A third view of the relation between environment and behavior appears to make the distinction meaningless by so defining the terms that environment and the individual become inseparable. According to this view, mankind and his habitat are indistinguishable except as aspects of the same entity. But existential and epistemological issues aside, these three and other views of the relation merge and blur when one attempts to answer questions such as, "How can one predict individual attributes at Time II from a knowledge of that individual's attributes and of the characteristics of the surrounding environment at Time I?" Stated that way, questions of causal priority may simply be ignored or attended to according to the tastes of the person answering the question. Elaborating the question, one might ask, "How can one predict individual attributes at Time III if between Times I and II the characteristics of the environment or the person or both changed in measurable ways?" Stating the questions in that manner brings us directly to a discussion of the concomitants of environmental change with respect to an aged population.

RESIDENTIAL MOBILITY, RELOCATION, AND ENVIRONMENTAL CHANGE

Let us turn to a review of the highlights of the literature pertaining to the relation between environmental change and the aging process. Minimal attention will be given to that body of literature dependent largely upon correlational analysis, in which some static aspects of the environment are associated with characteristics of the older person. It is our intention here to limit our attention to those studies in which the environment, or characteristics of it, have been allowed to vary.

The literature on environmental change in the field of gerontology falls naturally into two major categories. For obvious reasons, a significant proportion of that literature derives from studies conducted on institutionalized elderly samples. The remainder is based on studies of older people living in more or less conventional community settings.

While there is some reference in the literature to the studies of Camargo and Preston (1945), showing relatively high early mortality among elderly patients in hospitals for the mentally ill, one begins to pick up threads leading to current research thought in 1961, with Lieberman's study of the relation between mortality rates and entrance into a home for aged (Lieberman, 1961). Noting that the death rate during an 11-year period was three times higher during the first year of residence than the rate during the waiting period (four to six months), Lieberman attempted to compare survivors with those who had died in the first year of residence with respect to prior physical status, as determined by an examination of records. On the basis of the data, Lieberman was obliged to reject the notion that the increased mortality rate was highly associated with the physical status of those relocated at the time of their entrance into the institutions. The value of this study today, 15 years after, lies in its providing us with the rationale: "If it isn't due to differences in physical status, perhaps it is due to some aspect of the new environment."

Two years later, Aldrich and Mendkoff (1963) described a study based on the natural experiment consequent to the closing of the Chicago Home for Incurables, resulting in the relocation over a two-year period of 233 disabled persons, of whom 162 were 70 years of age or older. In that study Aldrich and Mendkoff asked three questions: "Does relocation affect the death rate of elderly and disabled persons? Does the anticipation of relocation affect the death rate? Does the patient's psychological adjustment affect his chances of survival?" We see at this stage again that, even thought the phenomenon being studied is human response in the face of environmental change, environmental concepts were not yet introduced as terms in the research questions. But note that another concept has been introduced. The second question, contingent on the finding that relocation may have negative consequences, suggests that the decline may have its origins in the anticipation of the relocation rather than in the relocation itself. Unfortunately, Aldrich and Mendkoff were not able to draw conclusions about the effect of anticipation or relocation with confidence because of the lack of statistical significance, but the data do hint at a negative consequence of anticipation. And again, mortality rates consequent to the move were shown to be higher than previous experience would have anticipated. Moreover, the data tended to confirm the idea that some elderly are more vulnerable than others. In this instance, the psychotic patients were especially vulnerable, whereas the nonpsychotics who expressed anger "were more likely to survive than patients who retreated from the conflict situation by regression or denial. . ."

By 1965, the problem had begun to be stated in terms of environmental change (Lieberman, 1965). Referring to a program of research entitled "Adaptation and Survival of the Aged under Stress," Lieberman now uses terms such as "the socio-physical environment of an old person" and "responses to environmental stress." It is as if up until this point one might hypothesize that there were negative consequences accruing to relocation, but now one might entertain the hypothesis that there were negative consequences to some aspect of the environment or to changes in aspects of the environment. In their "Factors in Environmental Change" Lieberman and Miller described a study of 45 women who had been relocated from one public institution to another (Miller and Lieberman, 1965). They had discovered that relatively healthy older people (both physically and psychologi-

cally) can deteriorate or exhibit maladaptive reactions to environmental change and that the negative reaction is not necessarily related
to previously determined low adaptive ability. Lieberman then goes
on to postulate three ways of viewing the relocation process:

> One view, stemming from the symbolic interaction position would con-
> sider the impact of institutionalization on the individual in terms of the
> personal meaning it has for him—it is traumatic or not because of its
> personally relevant symbolic meanings. . . A second view of the problem is
> expressed by the continuity–discontinuity hypothesis which suggests that
> extensive changes in the environment require considerable adaptive capac-
> ity because a new environment demands new behavior patterns. . . .thus
> the move to a new environment is stressful to the extent that it requires
> relinquishing old patterns and roles and adopting new ones. (Miller and
> Lieberman, 1965)

Lieberman notes further, however, that neither of these views necessarily requires that we attend to the specific attributes and characteristics of the new environment. It is, according to these views, only the
fact of change itself that is significant. It seemed at this point that
Lieberman was on the verge of expanding on the third viewpoint,
requiring that he attend to specific characteristics of the environment.
But that was to wait. In the meanwhile there would be other studies,
some demonstrating that relocation is harmful while others demonstrated that some relocation could be executed without harm to the
older person.

The next several years then produced a number of studies involving relocation of older people from the community to institutions or
from one institution to another. One of the principal objectives of
much of this research was to identify the characteristics or combination of characteristics of the older person that would predict successful
or unsuccessful adjustment to the relocation. Anderson (1967) for
example, found that instead of institutionalization, variation in the
amount and quality of interaction was found to explain changes in
self-esteem. She concluded that institutionalization does not necessarily have detrimental effects on older prople. Markus, Blenkner, Bloom,
and Downs (1971) reported a study in two homes for the aged in
which they discovered that age, sex, and residency experience alone
were not efficient predictors of postrelocation mortality. Then, in a
later paper (Markus, Blenkner, Bloom, and Downs, 1972) they tested
the predictive value of three major variables: Perceptual field-depend-
ence/independence, mental status, and physical status (containing
seven indices). Of particular lasting interest in this set of studies is the
discovery that predictable differences in one of the institutions could
not be replicated in the other institution. With the advantage of

hindsight, it may be worth noting for this review that conspicuous by their absence were measures of environmental characteristics that may have differentiated between the two institutions.

Markson and Cummings (1974, 1975), in studying the relocation of patients from mental hospitals in New York State, concluded that they could find no evidence that "forced relocation of a group of relatively physically healthy patients from one institution to another is sufficiently stressful to have any impact on their mortality experience." On the other hand, Bourestom and Tars (1974) reported the longitudinal assessment of the consequences of involuntary relocation of elderly patients from two medical facilities and compared that with the patients in a third facility in another state who were not relocated, and were again able to detect differences in mortality. What is of especial interest to this reviewer is that the two experimental settings could be differentiated on the basis of the degree of environmental change. They tell us:

> In one of the Michigan facilities patients were to undergo a radical environmental change from the county facility to a new and much larger proprietary nursing home in a nearby community. For these patients the change was total and adjustment had to be made to a new staff, a new program, a new physical environment, and new patient population. By contrast, patients in the other Michigan facility had many fewer adjustments to make; although the move was also involuntary, these patients experienced a moderate change in their physical environment only, namely a move to a new building several hundred yards away.

According to Bourestom and Tars, the results show:

> . . .that more radical environmental changes are associated both with higher mortality rates and with more negative changes in like patterns. . .the poor adjustment of the survivors of the radical relocation in this study indicate that the destructive effects of the environmental change are not limited to higher mortality rates.

Pastalan (1973) reporting on the same experiment, elaborates somewhat. According to him, "the best predictor of mortality is health status and cognitive function . . . home range as an environmental variable shows early promise of being a strong predictor of mortality." Home range is defined by Pastalan as "the total number of behavior settings penetrated by a patient during a typical day." According to him, "those patients who had ten or more behavior settings penetrations during a given day all survived. Those who had seven or fewer were highly vulnerable." While as strong a case could not be made for the predictive value of spatial autonomy, privacy preferences, and environmental coping style, Pastalan is optimistic that they may eventually be demonstrated to be effective predictors of improvement or deterioration, if not mortality. Now, over a 20-year period, what appears to have happened is that the initial awareness of increased mortality as a function of institutional relocation, with emphasis on the two words "relocation" and "mortality," has been transformed into the study of the detrimental effects of institutional environmental change. That pursuit has been proliferated by an interest in the study of the personal attributes of the patient that may enhance or inhibit the detrimental effect, and to a lesser extent, an interest in the definition of the characteristics of the environment that, when changed, can contribute to the decline of the patient's well-being.

Finally, we return to Morton Lieberman (1974) who, in a discussion on relocation research and social policy, summarizes just how far he (and therefore in some respects, we) have come. By this time Lieberman, who was present at the beginning, so to speak, was now prepared to recognize the "overpreoccupation" with death rates. "A more useful orientation for studying the effects of the relocation," he tells us, "would be to ask the question whether the individual departs, subsequent to being moved, in major ways from his prior physiological behavioral and psychological status." Those concerned with health care delivery and the provision of service have an especial interest in the debate regarding the appropriate outcome variable, but it may be assumed that the members of the design professions, who are primarily concerned with effecting environmental change, will find a more moderate criterion than death rate to be a more comfortable basis for evaluating the kinds of changes in environment they are able to effect. In summarizing his work regarding environmental considerations for social policy, Lieberman establishes a hierarchy for a set of priorities. "Clearly," he says, "the maintenance of the elderly in supportive community environments which minimize relocation is the best strategy but often is not feasible." But lower on the list of priorities would be those changes in environment that would by their nature require changes in their life-style of the elderly relocatee. Somewhere in between one would recommend those environmental changes that in themselves do not demand significant change in life-style. That environment characteristics do have a significant effect on adaptation to new surroundings, consequent to institutional relocation, is now sufficiently well documented to be accepted as fact. In summarizing his four studies, Lieberman is able to conclude that:

> . . .facilitative environments were those characterized by relatively high degrees of autonomy fostering, personalization of the patients, and community integration. Critical was that facilitative environment placed the locus of control much more in the hands of the patient, differentiating

among them and permitted them a modicum of privacy. Also, the boundaries between the institution and the larger community were more permeable than nonfacilitative environment. . . It is also important to point out that some environmental characteristics appeared to be not as crucial as might have been expected. For instance, the resource richness of the environment seemed to have little to do with the facilitation. (Lieberman, 1974)

If Lieberman's findings in this regard can be shown to be generalizable and replicable, we will have come a long way. It may appear to be unnecessarily contentious if it is pointed out that the definition of environmental attributes that contribute to "autonomy fostering", "personalization of the patients," and "community integration" might still be subject to some debate. In a real situation the definitions may be confirmed only after the fact of relocation. That is hardly a criticism of the work to date, but rather a hint of the mandate to researchers in doubt of the direction to be taken in the future.

We turn our attention now to a discussion of literature pertaining to environmental change and its impact on noninstitutionalized elderly. An historical approach may be useful here, also. While not addressed primarily to the concerns of the elderly, the work of Mark Fried on the West End of Boston may be a useful starting place (Fried, 1963). Paralleling the history of environmental research on the institutionalized elderly, the initial concern in the research on the noninstitutionalized elderly appears to be with the effects of relocation rather than the effects of environmental characteristics. Fried studied nearly 600 residents of a working-class neighborhood in Boston who were being forced to relocate because of urban renewal. While the sample was not predominantly elderly, we may assume that many of those who were relocated were retired. Fried found that significant proportions of those who were forced to relocate showed marked evidence of grief in relation to the move and that this response was extremely frequent among those who had previously said that they liked living in the area from which they were relocated. (Fully 75% of those who liked the area "very much" exhibited severe grief reactions even two years after the move.)

Fried was not particularly interested in the study of environment per se and consequently ought not to be faulted for neglecting to include some measures of the characteristics of the "lost home" and of the new environment to which the relocatee was forced to migrate. But, nonetheless, the study has utility for our analysis inasmuch as it demonstrated at an early date that, to the extent that environmental change is confounded with relocation or migration, the "pure" effects of the former may be determined only after the effects of the latter are accounted for. We shall return to that point later. The Fried study also contains the beginnings of another avenue of investigation in this field; the concern for the distinction between voluntary and involuntary relocation. It is to be noted that the relocated residents of Boston's West End were, by and large, reluctant to move. Another study, several years later (Terreberry, 1967) demonstrated similar findings again on a nonaged population, in which four out of five involuntary relocatees believed that relocation had a negative effect on their lives. About 40% of all those interviewed felt that the sense of loss had remained even until the time of the interview, which was anywhere from one to three years after the move.

The two studies just alluded to did not address themselves specifically to the problems of the elderly. However, at about this time another investigation was undertaken to determine the effect of "involuntary relocation because of interstate highway construction on the personal and social adjustment of a sample of elderly residents located in a western city of approximately 250,000 inhabitants [Kasteler, Gray, and Carruth, 1968]." Although in this study the "before" measures are missing, a sample of elderly involuntary relocatees (aged 55 and older) was compared with a comparable sample of nonrelocatees. The analysis demonstrated that those who had been relocated scored relatively "worse" than the nonrelocated with respect to adjustment on such dimensions as present health, intimate contacts, security, religious activity, and total personal adjustments. A similar negative impact of involuntary relocation was indicated with respect to a set of attitudes having to do with health, friends, work, economic security, family, happiness, and a total attitude score. The authors reached the general conclusion that "involuntary relocation was an especially stressful experience for older persons whose ties are generally more firmly established and who, by the process of aging, may be more resistant to change than are younger persons." Issues of experimental design aside, the study is useful in confirming implied hypotheses regarding the negative impact of involuntary change among the elderly. But what of voluntary relocation?

We turn now to what may be considered a landmark study in the gerontological literature pertaining to environmental change. In 1966, Carp published the results of a project begun years earlier to determine the effects of relocating into a high-rise public housing facility—Victoria Plaza—especially designed for the elderly (Carp, 1966). The publication of the study, long awaited by other investigators, brought to the attention of the field a wealth of information pertaining not only to the design of the study, but to findings with respect to changes in life-style and adjustment, social processes, and the prediction of tenant morale. But in the context of the present review, the study is of particular significance because those who moved to Victoria Plaza did so, by and large, voluntarily. Moreover, objectively Victoria Plaza was qualitatively better than the previous environment from which the tenants had moved. Carp, in a later paper, summarizes the findings in the following terms:

> Evidence of the dramatic effect of improved life-setting on this group of older people was overwhelming and was similar for men and women. . . . Consistently, scores of residents improved, those of non-residents showed no change or slight decrement. . . . The environmental change was deeply and almost unanimously satisfying to people who lived in Victoria Plaza. . . . They had fewer health complaints, and, among those they had, fewer were neurotic in type. They spent less time in bed on account of illness and less time in health care activities. . . . Unsuccessful applicants, initially similar to successful ones, exhibited little difference in behavior or attitude. . . (Carp, 1967, pp 106, 107)

It should be noted, in considering the Carp study, that three possible sources of influence on the findings cannot be separated: The moves were voluntary, the environmental change represented an improvement, and environmental change is confounded with relocation. Nonetheless, we cannot overlook the conclusions to which Carp draws our attention. Whereas earlier studies on involuntary relocation, and on institutionalization (which might be assumed to be involuntary) predominantly showed the negative consequences of relocation/environmental change, most of Carp's movers, in contrast, appear to improve.

Lawton and Yaffe (1970, comparing voluntary movers into a new apartment building with a comparable sample of nonmovers, found that after a year's time the movers were more likely to have improved in health but also more likely than the nonmovers to have declined. They suggest, modestly, that the new environment may, at least in some cases, facilitate rather than impede adaptive behavior. Subsequently, Lawton and Cohen (1974) attempting to answer the question, "Was the success story of Victoria Plaza the one-occasion result of an unduplicatable team, or is there reason to feel that the basic idea lends itself to replication by other teams?", assessed the experience of new tenants in five new housing sites, including low income housing, public housing, lower middle income housing sponsored by a fraternal organization, and a couple of 202 direct loan projects, one of which was sponsored by a progressive black home for the aged. Control groups of community residents were selected for comparative purposes. Multiple regression analysis was performed in such a way that the contribution to the variance of the outcome variable by the background demographic variables and premove scores, was removed prior to determining the contribution to variance by the critical independent variable, rehoused versus community resident. Lawton and Cohen conclude:

> With the exception of the health variable, the finding of this study over five housing environments is very consistent with Carp's findings at Victoria Plaza. . . It nonetheless seems evident that the effect was less marked in the present research.

They then go on to offer some tentative explanations for the difference in magnitude:

> In conclusion, it is not difficult to make a strong case for the overall favorable impact of new housing on elderly tenants. This conclusion must necessarily be tempered by the apparent greater decline in functional health among the rehoused than among controls.

Until recently, "environmental" research on community-based elderly has paralleled research on institutionalized elderly, adding the dimension of voluntariness to the analysis. The emphasis has been on relocation, with change in environment defined in only the most general qualitative terms. More recently, however, there is evidence in the literature that additional classes of variables are being introduced to the analysis. First, one need not labor the point that, while the involuntary relocation might result in decline and voluntary relocation might result in improvement, in any given sample of older people the effects are not universal. Recognizing the variability in response, Storandt, Wittels, and Botwinick (1975), for example, undertook a study to determine whether some of the factors predicting decline among institutionalized aged (e.g., Miller and Lieberman, 1965; Aldrich and Mendkoff, 1963) have predictive value with respect to the adjustment of relatively healthy old people relocating into noninstutional facilities. Specifically, at about the time of their move into a high-rise apartment building for the elderly, a number of aged subjects were assessed with respect to cognitive, personality, health, and activity dimensions. Approximately a year to a year and a half later the well-being of the elderly subjects was reassessed. A general conclusion offered by Storandt et al. is that

> . . .of the measures taken near the time of moving into an apartment complex for older adults, those of cognitive and psychomotor functions— rather than of personality, health, or activity—were predictive of well-being approximately 15 months afterwards. . . The predictors of poor

adjustment in the relocation of the institutionalized elderly were not seen here as predictors in the present sample of healthy older adults.

Carp, too, (1966, 1974) raises similar questions about differential adaptation. In the later paper she attempts to determine the degree to which a set of predictors, useful for predicting adjustment over an 18-month period among the new residents of Victoria Plaza, would also have been useful in predicting long-term adjustments over an eight-year period. In the present context what is important about these studies is not the specific identification of personal attributes and characteristics that are predictive of good adjustment, but rather the recognition that the effect of environmental change, whatever that might be, will be differential in magnitude and direction in accordance with that complex of personal attributes and characteristics.

A second recent variation on the mover/nonmover and voluntary/involuntary line of investigation has been the identification (at long last) of varying characteristics of the environment that may have differential effects. An illustration of this line of research is Lawton, Nahemow, and Teaff (1975). They examined the multivariate relations between public housing sponsorship, size of community, building size, and building height, on the one hand, to a set of measures of well-being, on the other. The authors summarize:

> Associations were found between both community size and non-profit sponsorship and some indices of well-being was found only in the case of housing satisfaction, when total project size was considered. On the other hand, tenants in low buildings were more satisfied with their housing and more motile in their environment than tenants in high-rise buildings with many personal and some other environmental factors controlled.

Finally, research by this author (Schooler, 1972, 1973, 1975) has attempted to separate the previously confounded effect of environmental change from the effect of mobility itself. In 1968 a project was undertaken involving interviews with approximately 4,000 older people who had been selected on an area probability basis from the noninstutionalized older population of the United States (excluding Hawaii and Alaska). That interview obtained information pertaining to health, morale, social relations, and environmental circumstances of the respondents. In addition, of course, a wide variety of demographic characteristics were ascertained. Each of the four principal domains was factor analyzed and factor scores for each individual response were computed. Those factor scores served as principal variables for the analysis, most of which took the form of cross-tabulation among environmental characteristics, social relations, health, and morale.

Without attempting to summarize all of the analysis of the 1968 data, one might say, in a nutshell, that it had demonstrated that some variation in health and morale may be associated with, and indeed attributed to, variation in environmental characteristics; that the effect of variation in frequency of social interaction is not consistent; and that variations in relationships can be associated with or attributed to variations in demographic attributes, even though in many instances the nature and direction of those associations defy simple explanation. Bearing in mind that we are describing at this point the implication of findings based solely on cross-sectional or one-point-in-time data, it seemed justifiable on the basis of the 1968 data alone to draw conclusions about the effects of environment, in the sense that differential adaptation to environmental attributes could be defined.

In one of the later papers to come out of that project (Schooler, 1973), the author pointed out that because environmental change usually means that the person moves from one environment to another, it must be recognized that the effect of the change in environment is necessarily confounded with the effects due to mobility. The work of other investigators, principally Lieberman and his associates, referred to earlier in this chapter, suggests that the elderly in particular are frequently seen to decline in health and morale when they relocate, for example, from the community to a nursing home or from one nursing home or hospital to another, or from some institution into the community. Furthermore, it has been noted, at least on some occasions, that the anticipation of relocation is itself an apparent source of decline. In an attempt to determine the effect of environmental change and, it was hoped, to be able to separate those effects due to environment from those due to mobility, the 1968 study just summarized was elaborated by means of a reinterview three years later of a subsample of 521 cases. Sampling was conducted in such a way as to maximize the number of movers obtained.

Earlier research would have led us to believe that the anticipation of a residential move would lead to a decline in either health or morale or both. If such a hypothesis were warranted, one might expect to observe a decline in health and morale in 1971 more frequently among those who expected in 1968 to move than those who did not expect to move. The analysis of the 1971 data showed that, with respect to three out of four measures of morale, a greater proportion of those who expected to move than those who did not expect to move showed a decline in morale during the intervening three years. The fourth measure of morale, for reasons yet undetermined, showed a small but significant reversal of that direction. Similarly, of the three measures of health used, the first and third, which were measures of disability, showed significant reversals of the predicted direction of the relationship. On the other hand, the second measure of health, which was a measure of general health, showed a small but very significant relation in the predicted direction. So much for expectation; sometimes it appears to result in decline and sometimes it doesn't.

What about moving? The analysis of Schooler's 1971 data demonstrates that changing one's residence appears to have a detrimental effect in two out of the four morale factors and in two out of the three health factors. With respect to the remaining factors, the results would appear to be equivocal. At that point in the analysis, then, the findings of the study were in many instances in the directions initially predicted. But if anticipation of moving did not *always* result in decline and if actual moving did not *always* result in decline, was it not likely that some form of interaction was taking place? Indeed, that was the case. Schooler finds that with the exception of the case of one morale factor, those who show the greatest incidence of decline in health and morale are those who both expected to move in 1968 and have moved by 1971. Schooler argues at this point that the anticipation of residential relocation frequently initiates a decline in health or morale that is reinforced or accelerated by an actual move. He raises the question, however, whether it may be possible to mitigate the negative effect of this interaction if the new environment is in some sense supportive. By the same token, might it not be expected that if the characteristics of the new environment are less supportive than those of the original environment, even greater deterioration may take place? To test this proposition, he isolated those elderly respondents who not only anticipated in 1968 that they would move, but in fact had moved by 1971. In addition, the characteristics of the 1971 environment were compared with the characteristics of the 1968 environment to determine whether or not those environmental characteristics had changed. In that analysis, Schooler employed the measure called "age-related morale." In the case of three measures of change in environmental characteristics (quality of dwelling, convenience of service, and availability of social services) Schooler is able to show a striking association in the predicted direction between change in morale and change in environment. That is, among those who both expected to move and did move (i.e., were very vulnerable), morale declined most among those who experienced a deterioration in environment.

In the most recent paper (Schooler, 1975) the analysis is further elaborated. First, the difficulty of disentangling the confounded affects of mobility and environmental change was demonstrated by showing that, in the aggregate, the quality of residential environment is more likely to change for the worse among those older people who do not move than among those who do move!

> Thus, if changes in morale or health were to be associated with moving, without control for the change in the environment, or if similarly the simple bivariate relation between environmental change. . .and change in the dependent variable were to be shown, we might expect the negative effect of moving to be offset to some extent by the positive effect of improved environment, resulting in spuriously low relationships. (Schooler, 1975, p. 164)

Then Schooler shows that the relation between mobility and change in morale and health (employing four measures of morale and the measure of general health) is in each instance quite low (gamma ranging from .006 to .251). But, when the relation is controlled for qualitative change in the environment, in all but one instance the magnitude of the relation between mobility and the dependent variable increases noticeably. Reversing the procedure, that is, controlling the relation between the outcome variables (morale and health) and environmental change by residential mobility, similar if not equally impressive results were obtained. Schooler concludes from this analysis that the change in environment can be separated from residential mobility and the effects of each separately determined. For theoretical reasons to be discussed in a later segment of this chapter, Schooler then draws the distinction between a change in environment that includes the notion of direction of change (improvement, decline) and change in the environment that is in essence nondirectional. In other words, the question raised is whether it is merely the change (in contrast to stability) in the environment, or is it the qualitative aspect of the change that is responsible for the effect? The analysis of this portion of the investigation shows that among movers there are very few statistically significant relations between nondirectional change in the environment and change in the dependent variable. On the other hand, among nonmovers, most of the relations are significant. Schooler makes little more of this finding than to suggest that nondirectional change is a meaningful concept.

What can be made of this cursory examination of the literature of environmental change and its impact on older people? The design of most of the research reported here does not permit one to speak separately of the effect of environmental change and the effect of relocation. While the Schooler data would suggest that most of the variation is due to relocation, a smaller but undoubtedly significant portion of the variation must be due to qualitative changes in the environment itself. But even if one were to attribute all of the effects to environmental change, it must be noted that the specific attributes of the environment involved in the change, with few exceptions, have not been systematically delineated. The problem, however, goes beyond the obvious absence of labels for environmental characteristics. There is the taxonomic problem, to be sure: At the present time, we have no complete, widely accepted, classificatory scheme for all the environmental descriptors. But even if we did, we would be hard pressed in studies comparable to those described here to disentangle the effects of environmental characteristics that occur jointly. (That is, to use a highly oversimplified and trivial example, a change in a physical attribute, such as level of illumination, may be accompanied by a change in another physical attribute, say, total area.) But even if we had a taxonomy that was generally accepted and even if we could overcome the methodological difficulty of separating effects, we would be faced with yet another problem, which is the delineation of the conceptual linkage between the taxonomic categories and some larger theory accounting for the association between environmental change and response of the elderly. It is in recognition of that latter problem that this discussion now turns to a description of current theory.

CURRENT THEORY

The preceding cursory review of literature pertaining to the effect of environmental change informs us as follows: The status of the individual involved in environmental change may improve or may decline, depending on a multitude of other characteristics of the situation. In addition, we may surmise that environmental change has two facets: One facet is mobility while the other is alteration of the attributes of the environment. In particular situations, either of these dimensions may have a value of zero, yet both may be operating separately. Clearly, what is needed at this particular point is a theory of the middle range that might give order to the meager findings we possess while at the same time giving direction to future research.

In the field of gerontology a handful of attempts at such theorizing are available to us and will be described here. One notable significant attempt comes from Lawton and his collaborators in a series of papers culminating in "Competence, Environmental Press, and the Adaptation of Older People"(Lawton, 1975). Beginning with the familiar equation, $B = f(P, E)$, Lawton develops in somewhat greater detail the terms of that ecological equation that will be useful in his theory. For Lawton, the most relevant aspects of P (Person) for his theory are subsumed under the term *competence*. Competence is manifested in biological health, sensory-perceptual capacity, motor skills, cognitive capacity, and ego strength. Competence then is conceptually defined by Lawton as "the theoretical upper limits of capacity of the individual to function in the areas of biological health, sensation-perception, motoric behavior, and cognition." Turning his attention to E (Environment), Lawton then focuses on Murray's concept of *press* (Murray, 1938). Murray, according to Lawton, used *press* to refer to an environmental force that tended to activate an interpersonal need, while he, with Nahemow, developed a model in which the concept of need was not used. "Instead, press are defined in normative terms. That is, an environmental stimulus or context is seen as having potential demand character for any individual if empirical evidence exists to demonstrate its association with a particular behavioral outcome for any group of individuals [Lawton, 1975]." Lawton follows this discussion with an elaboration of the essential equation by including the interaction term, $P \times E$:

$$B = f(P, E, P \times E)$$

Lawton and Nahemow (1973) have developed a graphic representation (Figure 1) that allows the reader to more easily grasp the implications of their theoretical framework from this point on. Environmental press is represented as a continuum on the abscissa. Competence is represented also as a continuum on the ordinate. It goes without saying that such a two-dimensional representation sacrifices much in order to achieve simplicity in the explanation. But it stands to reason that what appears in the diagram to be unidimensional is, of course, a multidimensional complex of attributes and variables, in the case of environment on the one hand and in the case of the individual on the other. Oversimplification of reality notwithstanding, the diagram does have utility. Note that at least conceptually an individual could be located in that two-dimensional space at a given point in time with respect to both his competence and the environmental mileu. Note further that if a horizontal line is drawn, representing a given level of competence, both affect and behavior at that point in time will, according to the scheme, be determined by the strength of environmental press. Likewise, for a given level of press, represented by a vertical line, affect and behavior are inferred to be a function of level of competence. Examination of the diagram reveals further that for any given level of competence, as the stimuli in the environment intensify, that is, as environmental press strengthens,

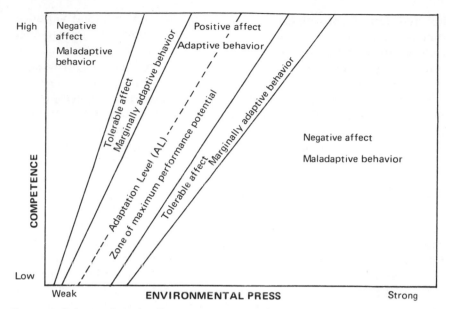

Figure 1. Behavioral and affective outcomes of person–environment transactions. (Adapted from Lawton & Nahemow, 1973.)

behavior will become maladaptive and affect will become negative. Lawton also incorporates the concept of adaptation (Helson, 1954), represented by the line drawn through the middle of the inverted triangle in the diagram. According to Lawton, "Not only is there a tendency for every individual experiencing a given environmental press to establish an adaptation level, but the magnitude of that neutral stimulus level is partly determined by the competence of the individual." According to the diagram, further, "the lower the competence of the individual, the lower the strength of press must be in order to maintain this steady state of automatic but adaptive behavior and neutral affect." Lawton elaborated his model even further by adding the idea of "the optimization principle," following the lead by Wohlwill (1974). According to Wohlwill, any minor deviation from the adaptation level in either direction, positive or negative, will result initially in positive affect but as the discrepancy from the adaptation level increases the experience of positive affect at some point becomes negative.

All of the preceding description is consistent with, or stems from, Lawton's earlier enunciated "environmental docility hypothesis" (Lawton, 1970). He summarizes that hypothesis as suggesting that lowered competence increases the proportion of behavioral variance that is associated with the environmental as compared to personal factors. In elaborating his theory, Lawton is as mindful of the implications of his model for change in competence as he is of changes in external environment. It goes without saying that this brief description hardly does justice to Lawton's complete theoretical development, but the theory's implication for the study of environmental change as it pertains to the elderly should be obvious. In the first place, as suggested by the description earlier in this chapter of the deficits associated with the aging process, the translation of "deficit" to "level of competence" and the likelihood that change in environment can be conceptualized in terms of change in level of press, suggest immediately the utility of the theory and warrants continued investigation. Lawton himself has demonstrated this by interpreting the findings of the studies of others in the context of his own theoretical statement. Of particular interest in the study of environmental change in Lawton's demonstration that studies of Aldrich and Mendkoff (1963), of Goldfarb, Shahinian, and Turner (1966), of Markus (1970), or Killian (1970), and of Markus, Blenkner, Bloom, and Downs (1970) among others can be interpreted in a manner consistent with the model. That more work is needed will become apparent, but Lawton himself provides help in suggesting the present inadequacies of his model. The conceptual, as well as methodological, problems associated with the more complex aspects of competence are alluded to. With respect to environmental press, Lawton recognizes the need of a functional taxonomy and the difficulties of operationalizing some of the concepts. Other problems associated with the concept of adaptation level and personality style are noted.

A second major contribution to theory in the gerontological literature has been developed by Eva Kahana (1975). She refers to her statement as a congruence model of person–environment fit. While there appears to be some similarity between Kahana's and Lawton's theories, they differ in some significant respects. She, like Lawton, traces the roots of her theory to Murray's model involving need and press. But, whereas Lawton transformed Murray's concern for need into his own use of the concept of competence, Kahana is satisfied to continue with the conception of personal need. Accordingly, in her model, if personal needs or preferences are congruent with the environmental characteristics a positive outcome ensues. But, where needs are dissonant with characteristics of the environment, one might expect either negative outcome or (invoking Helson's adaptation level) adaptation might result, leading further to positive outcomes. Drawing on the work of Kleemeier, Kahana notes three dimensions of environmental differences: the segregate dimension, the congregate dimension, and the institutional-control dimension.

a. The *segregate dimension* was originally used by Kleemeier (1961) to refer to "the condition under which older persons may live exclusively among their age peers having little contact with other age groups."
b. The *congregate dimension* refers to "the closeness of individuals to each other and to the degree of privacy possible to attain in the settings."
c. The *institutional-control dimension* refers to the extent of staff control of residents, the use and importance of rules and the degrees of resident autonomy which are tolerated. (Kahana, 1975)

In addition to these three dimensions, Kahana notes three dimensions of congruence based on characteristics of the aged individual. These are: the need for activity versus passivity, the affective expression of needs, tolerance for ambiguity, and impulse control. These seven dimensions are further elaborated, as seen in Table I. Note that for each subdimension there is both an environmental and individual set of adjectives. The theory in its broadest terms predicts that outcome will tend to be positive to the extent that the needs and preferences represented on the right-hand side of the table are congruent with the characteristics of the environment listed on the left-hand side. Kahana has provided some test of her model through a study in a variety of nursing homes and homes for the aged settings. Her findings indicate that morale of the older people can be predicted or explained by five of the subdimensions of congruence. As she describes it:

Privacy was among the best predictors of morale in all three homes and motor control, stimulation, continuity with the past, and change vs. saneness appeared as best predictors in two of the three homes. The occurrence of these five subdimensions as important predictors in at least two homes provides fairly firm grounds for attesting the salients of these congruent subdimensions in predicting morale. (Kahana, 1975)

As in the case of the previous description of Lawton's model, the brief overview presented here does not do justice to the complete presentation of Kahana's position. In anticipation of readers' objections, it may be noted that Kahana is aware of and discusses the various variations of her basic model, addressing such issues as: the difference between a conception of dissonance that takes directionality into account and one that does not; or a model that attends to the specific area in which dissonance occurs as opposed to one focusing simply on cumulative dissonance. It is obvious that Kahana's model attempts to account for outcome (morale, behavior) in a static situation in terms of congruence, as opposed to a model that accounts for

TABLE I
DIMENSIONS OF CONGRUENCE

Environment	Individual

1. Segregate Dimension

A. Homogeneity of composition of environment. Segregation based on similarity of resident characteristics (sex, age, physical functioning and mental status).

A. Preference for homogeneity, i.e., for associating with like individuals. Being with people similar to yourself.

B. Change vs. sameness. Presence of daily and other routines, frequency of changes in staff and other environmental characteristics.

B. Preference for change vs. sameness in daily routines, activities.

C. Continuity or similarity with previous environment of resident.

C. Need for continuity with the past.

2. Congregate Dimension

A. Extent to which privacy is available in setting.

A. Need for privacy.

B. Collective vs. individual treatment. The extent to which residents are treated alike. Availability of choices of food, clothing, etc. Opportunity to express unique individual characteristics.

B. Need for individual expression and idiosyncrasy. Choosing individualized treatment whether that treatment is socially defined as "good" treatment or not.

C. The extent to which residents do things alone or with others.

C. Preference for doing things alone vs. with others.

3. Institutional Control

A. Control over behavior and resources. The extent to which staff exercise control over resources.

A. Preference for (individual) autonomy vs. for being controlled.

B. Amount of deviance tolerated. Sanctions for deviance.

B. Need to conform.

C. Degree to which dependency is encouraged and dependency needs are met.

C. Dependence on others. Seek-support, nurturance vs. feeling self-sufficient.

4. Structure

A. Ambiguity vs. specification of expectations. Role ambiguity or role clarity, e.g., rules learned from other residents.

A. Tolerance of ambiguity vs. need for structure.

B. Order vs. disorder.

B. Need for order and organization.

5. Stimulation—Engagement

A. Environment input (stimulus properties of physical and social environment); (not only availability of stimulation even to which that is directed to resident).

A. Tolerances and preference for environmental stimulation.

TABLE I. *continued*

Environment	Individual
B. The extent to which resident is actually stimulated and encouraged to be active. (Jackson)	B. Preference for activities vs. disengagement.

6.   *Affect*

A. Tolerance for or encouragement of affective expression. Provision of ritualized show of emotion (e.g., funerals).	A. Need for emotional expression-display of feelings, whether positive or negative.
B. Amount of affective stimulation. Excitement vs. peacefulness in environment.	B. Intensity of affect, e.g., need for vs. avoidance of conflict and excitement (shallow affect).

7.   *Impulse Control*

A. Acceptance of impulse life vs. sanctions against it. The extent to which the environment gratifies needs immediately vs. postponed need gratification. Gratification/ deprivation ratio.	A. Ability to delay need gratification. Preference for immediate vs. delayed reward. Degree of impulse need.
B. Tolerance of motor expression— restlessness, walking around in activities or at night.	B. Motor control; psychomotor inhibition.
C. Premium placed by environment on levelheadedness and deliberation.	C. Impulsive closure vs. deliberate closure.

* Adapted from Kahana, E. A congruence model of personenvironment interaction. In P. G. Windley, T. O. Byerts, and F. G. Ernst (Eds.), *Theory development in environment and aging*, Washington, D.C.. The Gerontological Society, 1975

differential outcome as a consequence of change. But, granted the basic credibility of the model for static situations, it would take little effort to modify it in order to account for environmental change. Thus, for example, we could characterize change as being in the direction of congruence or having the effect of disrupting congruence.

A third theoretical framework useful for interpreting the relation between environmental change and aging is one proposed by this author (Schooler, 1975). It is presented as a stress-theoretical point of view, and should be thought of as a restatement or interpretation of the theoretical point of view of Richard Lazarus (Lazarus, 1966). Indeed, when this point of view was first presented, it was stated as a deliberate effort to attempt to apply existing theory to current data rather than to attempt to develop a new theory, *de novo*. The theoretical statement was developed in order to explain observed variations in adaptive responses of the elderly, specifically in terms of health and morale. The statement was formulated to attempt to account for the impact of characteristics of the physical environment and changes in the environment. In order to do so, the concept of stress, mediated by cognitive processes between environment on the one hand and the adaptive responses on the other was imposed on the theory. Some of the previously noted observations to be accounted for by the theory were: (1) Some environmental attributes are associated with negative states of morale and health. (2) Mobility resulting in a negative change in environment results in negative changes in morale and health. (3) Residential mobility, even when not accompanied by perceived changes in the quality of environment, results in the decline of morale and health. (4) Anticipation of mobility results in decline in morale and health. (5) When risks are cumulative, that is when movement is anticipated and it does in fact occur, the greatest deterioration in morale and health is noted. (6) Many of the negative effects noted above appear to be buffered by the presence of certain kinds of social relations, specifically those that appear to sustain self-esteem. (7) The preceding statements of relationships are stated as generalizations, but not all combinations of environmental characteristics and outcome measures can be shown to conform to those generalizations.

Lazarus introduced his theory of psychological stress by noting:

> Psychological stress analysis . . . is distinguished from other types of stress analysis by the intervening variable of *threat*. Threat implies a state in which the individual anticipates a confrontation with a harmful condition of some sort. Stimuli resulting in threat or non-threat reactions are cues that signify to the individual some future condition, harmful, benign, or beneficial. These and other cues are evaluated by the cognitive process of *appraisal*. The process of appraisal depends on two classes of antecedents. The first class consists of *factors in the stimulus configuration*, such as the comparative power of the harm producing condition and the individual's counter harm resources, the imminence of the harmful confrontation, and degree of ambiguity in the significance of the stimulus cue. The second class of antecedents that determine the appraisal consists of *factors within the psychological structure of the individual*, including motive strength, and pattern, general beliefs about transactions with the environment, intellectual resources, education, and knowledge . . . Once the stimulus has been appraised as threatening, processes whose function it is to reduce or eliminate the anticipated harm are set in motion. They are called *coping processes*. They also depend on cognitive activity but because they are influenced by special factors we term the cognitive activity related to coping *secondary appraisal*. Three main classes of factors are involved in secondary appraisal. The first is *degree of threat*. The second consists of *factors in the stimulus configuration* such as the locatability and character of

the agent of harm, the viability of alternative available routes or actions to prevent the harm and situational constraints which limit or encourage the action that may be taken. A third class of factors influencing the secondary appraisal process is *within the psychological structure*—for example the pattern of motivation which determines the price of certain coping alternatives, ego resources, defensive dispositions, and general beliefs about the environment and one's resources for dealing with it.

Secondary appraisal based upon the above factors determines the form of coping process, that is, the coping strategy adopted by the individual in attempting to master the danger. The end results observed in behavior (for example, affective experiences, motor manifestations, alterations in adaptive functioning, and physiological reactions) are understood in terms of these intervening coping processes. Each pattern of reaction—for example, action aimed at strengthening the individual's resources against the anticipated harm, attack with or without anger, anger without attack, avoidance with fear, fear without avoidance, and defense—is determined by a particular kind of appraisal. (pp. 24–26)

According to Lazarus, the appraisal of threat depends not only on the stimulus configuration itself (i.e., the environment), but on psychological characteristics of the individual. Thus, to understand the impact of environmental change according to this theoretical formulation, an understanding of cognitive process and the correlates or determinants of cognitive process is required. It is critical to note at this point that Lazarus includes the possibility of the importance of social factors as well as the psychological and biological at this particular point in his paradigm. Now, it would be useful to introduce a line of reasoning presented by John Cassel (mimeo). Using the spread of infectious disease as a function of crowding as the analogy, and nothing that that relationship is not always demonstrable, Cassel asserts:

> . . .in a large number of cases clinical manifestations of disease can occur through factors that disturb the balance between the existing ubiquitous organisms and the host that is harboring them. It may well be that under conditions of crowding, this balance may be disturbed, but this disturbance is not simply a function of the physical crowding (that is, the closeness of contact of susceptible to carriers of microorganisms) but of other processes.

Cassel's paper is concerned with the links between physical residential environment and health. He proposes that it is the presence of social relationships that may constitute one of the environmental factors that affect homeostatic mechanisms. Viewed in this manner, Cassel's theory then is compatible with the larger theory of stress outlined by Lazarus. That is, it is not only the psychological but also sociological phenomena that may determine the cognitive processes that allow environmental characteristics and environmental change to be appraised as threatening and therefore stressful. On the basis of a review of animal research, Cassel derived a number of principles, one of which is concerned with "the available protective factor, those devices which 'buffer' or 'cushion' the individual from the physiological or psychological consequences of the social processes." In addition to biological adaptive processes, Cassel suggests that some social processes have also been shown to be protective. In particular, he asserts, are the nature and strength of the group support provided to the individuals. He summarizes then as follows:

> . . .according, much of the work concerned with social or psychological antecedents to disease has attempted to identify a particular situational set (usually labeled "stress" or "a stressor") which would have a specific causal relationship between typhoid bacillus and typhoid fever. Such a formulation would appear to be clearly at variance with the animal data, a striking feature of which is the wide variety of pathologic conditions that emerged following changes in the social mileau. A conclusion more in accordance with the known evidence then would be that such variations in group relationships, rather than haveing a specific etiological role, would enhance susceptibility to disease in general. (Cassel, mimeo)

It is this author's view that Cassel's formulation is altogether compatible with the paradigm presented by Lazarus. If, according to Lazarus, environmental change may be appraised as threatening and if that appraisal is determined by the interaction of the stimulus configuration with attributes of the individual that themselves are psychological or social in origin, then it is in regard to this process between stimulus and stress that Cassel's concern for social (and also biological) buffers becomes relevant. In summary, the stress-theoretical formulation may be stated as follows: Environmental change may, through the cognitive process of *appraisal*, be interpreted as threatening. Whether or not such an appraisal will take place is a function, first, of the nature of that environmental change, but, secondly, it will be a function of factors within the psychological structure of the individual. Then, in the presence of threat, forces are mustered in order to cope with the situation. The processes of coping are also influenced by the cognitive activity Lazarus labeled "secondary appraisal." Following the lead of Cassel, both the primary and secondary appraisal processes themselves, it is contended, may be influenced by the buffering effect of essentially social processes and phenomena. The larger presentation (Schooler, 1975) then proceeds to demonstrate that the kinds of observations yielded by the data and alluded to earlier in this chapter, can be explained or accommodated within a stress-theoretical framework such as Lazarus's.

The three statements of theory described here are by no means exhaustive, but they are typical of the social–psychological–ecological thinking that characterizes the field of social gerontology at this time. Yet, while there are elements of difference, it may be useful to see if the terms of one theory can be translated into the terms of the other. It is obvious at the outset, for example, that environmental press is common to both the Lawton and Kahana formulations and need not be considered in any way different from the "stimulus configuration" concept of the Lazarus-Schooler formulation. But what can we make of the Kahana "congruence" term in relation to the Lawton "competence' term? It seems reasonable that an equation can be stated as follows: Competence, as in Lawton's categories of biological health, sensory-perceptual capacity, motor skills, cognitive capacity, and ego strength, equals the ability to satisfy preferences, as in the case of preference for homogeneity, or preference for change versus sameness, or to meet needs (such as the need for privacy or the need to conform). Competence may also be equated with Kahana's "tolerance" as in the case of tolerance for ambiguity and tolerance for environmental stimulation. In short, Lawton's competence may be thought of as the inherent, innate, or learned ability to meet the needs specified by Kahana's presentation. Conversely, Kahana's congruence can be achieved (other than fortuitously) by the person only if they are competent in Lawton's terms. Attempting to justify the two theories in this manner gives undue emphasis to the volitional aspect of the person–environment situation.

Since this chapter addresses itself to the concerns of the elderly population, a population at risk and subject to the deficits reported in the earlier part of the chapter, we should at least acknowledge that congruence (in Kahana's terms) can be achieved and competence (in Lawton's terms) not exceeded by judicious intervention in the environmental situation, that is, through planned environmental change. Moreover, both Lawton's and Kahana's theories are compatible with a stress-theoretical framework (Schooler) in the sense that dissonance (in Kahana's terms) might be appraised as threatening just as inappropriate levels of press for given levels of competence (in Lawton's terms) would be so appraised. But all three theories, whether employed in some synthesized form or whether treated as competitive to each other, have especial merit for the study of the aging process, largely because each is inherently able to acknowledge the increased vulnerability of the aging population and to relate that vulnerability to considerations of a changing environment.

ACKNOWLEDGMENTS

The assistance of E. Bryant and B. Holtzman and the specific contribution of R. Dunkle to the completion of this chapter are gratefully acknowledged.

REFERENCES

Aldrich, C. K., and Mendkoff, E. Relocation of the aged and disabled: A mortality study. *Journal of the American Geriatrics Society*, 1963, *11*, 185–194.

Anders, T. R., Fozard, T. L., and Lillyquist, T. D. The effects of age upon retrieval from short-term memory. *Developmental Psychology*, 1972, *6*, 214–217.

Anderson, N. N. Effects of institutionalization on self-esteem, *Journal of Gerontology*, 1967, *22*, 313–317.

Baltes, P. B., and Schaie, K. W. Aging and IQ: The myth of the twilight years. *Psychology Today*, 1974, *7*(10), 35–40.

Beattie, W. M., Jr. Discussion, in Aging in the year 2000: A look at the future. *The Gerontologist*, 1975, *15*(1), 39.

Botwinick, J. *Aging and behavior*. New York: Springer Publishing Co., Inc., 1973.

Bourestom, N., and Tars, S. Alterations in life patterns following nursing home relocation. *The Gerontologist*, 1974, *14*, 506–510.

Butler, R. N. *Why survive? Being old in America*. New York: Harper & Row, 1975.

Camargo, O., and Preston, G. H. What happens to patients who are hospitalized for the first time when over 65? *American Journal of Psychiatry*, 1945, *102*, 168–173.

Carp, F. M. *A future for the aged*. Austin: University of Texas Press, 1966.

Carp, F. M. The impact of environment on old people. *The Gerontologist*, 1967, *7*, 106–108.

Carp, F. M. Short-term and long-term prediction of adjustment to a new environment. *Journal of Gerontology*, 1974, *29*, 444–453.

Cassel, J. The relation of the urban environment to health: Toward a conceptual frame and a research strategy. (mimeo, no date).

Conestrari, R. E., Jr. Paced and self-paced learning in young and elderly adults. *Journal of Gerontology*, 1963, *18*, 165–168.

De Vries, H. A. Physiology of exercise and aging. In D. S. Woodruff and J. Birren (Eds.), *Aging*. New York: D. Van Nostrand Co., 1975.

Eisdorfer, C. Verbal learning and response time in the aged. *Journal of Genetic Psychology*, 1965, *197*, 15–22.

Fried, M. Grieving for a lost home: Psychological costs of relocation. In L. Duhl (Ed.), *The urban condition*. New York: Basic Books, 1963.

Goldfarb, A. I., Shahinian, S. P., and Turner, H. Death rates of relocated nursing home residents. Paper presented at the 17th annual meeting of Gerontological Society, New York, November, 1966.

Helson, H. *Adaption level theory*. New York: Harper & Row, 1974.

Kahana, E. A congruence model of person–environment interaction. In P. G. Windley, T. O. Byerts, and F. G. Ernst (Eds.), *Theory development in environment and aging.* Washington, D.C.: The Gerontological Society, 1975.

Kasteler, J. M., Gray, R. M., and Carruth, M. L. Involuntary relocation of the elderly. *The Gerontologist,* 1968, *8,* 276–279.

Killian, E. C. Effect of geriatric transfers on mortality rates. *Social Work,* 1970, *15,* 19–26.

Kleemeier, R. W. The use and measuring of time in special settings. In R. W. Kleemeier (Ed.), *Aging and leisure.* New York: Oxford University Press, 1961.

Lawton, M. P. Institutions for the aged: Theory, content, and methods for research. *The Gerontologist,* 1970, *10,* 305–312.

Lawton, M. P. Competence, environmental press and the adaptation of older people. In P. G. Windley, T. O. Byerts, and F. G. Ernst (Eds.), *Theory development in environment and aging,* Washington, D.C.: The Gerontological Society, 1975.

Lawton, M. P., and Cohen, J. The generality of housing impact on the well-being of older people. *Journal of Gerontology,* 1974, *29,* 194–204.

Lawton, M. P., and Nahemow, L. Ecology and the aging process. In C. Eisdorfer and M. P. Lawton (Eds.), *The psychology of adult development and aging.* Washington, D.C.: American Psychological Association, 1973.

Lawton, M. P., Nahemow, L., and Teaff, J. Housing characteristics and the well-being of elderly tenants in federally assisted housing. *Journal of Gerontology,* 1975, *30,* 601–607.

Lawton, M. P., and Yaffee, S. Mortality, morbidity, and voluntary change of residence by older people. *Journal of the American Geriatrics Society,* 1970, *18,* 823–831.

Lazarus, R. S. *Psychological stress and the coping process.* New York: McGraw-Hill Book Co., 1966.

Lieberman, M. A. Relationship of mortality rates to entrance to a home for the aged. *Geriatrics,* 1961, *16,* 515–519.

Lieberman, M. A. Psychological correlates of impending death: Some preliminary observations. *Journal of Gerontology,* 1965, *20,* 181–190.

Lieberman, M. A. Relocation research and social policy. *The Gerontologist,* 1974, *14,* 494–501.

Maddox, G., and Douglas, E. G. Aging and individual differences: A longitudinal analysis of social, psychological, and physiological indicators, *Journal of Gerontology,* 1974, *29,* 555–563.

Manney, J. D., Jr. *Aging in American society.* Ann Arbor: Institute of Gerontology, Michigan-Wayne State University, 1975.

Markson, E. W., and Cumming, J. H. A strategy of necessary mass transfer and its impact on patient mortality. *Journal of Gerontology,* 1974, *29,* 315–321.

Markson, E. W., and Cumming, J. H. The post-transfer fate of relocated mental patients in New York. *The Gerontologist,* 1975, *15,* 104–108.

Markus, E. Post-relocation mortality among institutionalized aged. Cleveland: Benjamin Rose Institute, 1970 (mimeo).

Markus, E., Blenkner, M., Bloom, J., and Downs, T. Relocation stress and the aged. *Interdisciplinary topics in gerontology,* Vol. 6, Basel: S. Karger, 1970.

Markus, E., Blenkner, M., Bloom, J., and Downs, T. The impact of relocation upon mortality rates of institutionalized aged persons. *Journal of Gerontology,* 1971, *26,* 537–541.

Markus, E., Blenkner, M., Bloom, J., and Downs, T. Some factors and their association with post-relocation mortality among institutionalized aged persons. *Journal of Gerontology,* 1972, *27,* 376–382.

Miller, D., and Lieberman, M. A. The relationships of affected state and adaptive capacity to reactions to stress. *Journal of Gerontology*, 1965, *20*, 492–497.

Murray, H. A. *Explorations in personality*. New York: Oxford Press, 1938.

Neugarten, B. L. (Ed.). Aging in the year 2000: A look at the future. *The Gerontologist*, 1975, *15*(1), part II.

Pastalan, L. A. Panelist's paper. Involuntary environmental relocation: Death and survival. In W. F. E. Preiser (Ed.), *Environmental design research*. Vol. 2. Stroudsburg, Pa.: Dowden, Hutchinson & Ross, Inc., 1973, p. 410.

Riegel, K., and Riegel, R. Development, drop and death. *Developmental Psychology*, 1972, *6*, 306–319.

Rosow, I. *Socialization to old age*. Berkeley: University of California Press, 1974.

Schooler, K. K. Effects of changes in residential environment. Paper presented at the 9th International Congress on Gerontology, Kiev, Russia, July, 1972.

Schooler, K. K. Some consequences of environmental change in an elderly sample. Paper presented at the American Psychological Association, Montreal, Quebec, 27 August 1973.

Schooler, K. K. Response of the elderly to environment: A stress-theoretical perspective. In P. G. Windley, T. O. Byerts, and F. G. Ernst (Eds.), *Theory development in environment and aging*. Washington, D.C.: The Gerontological Society, 1975.

Smith, A. Aging and interference with memory. *Journal of Gerontology*, 1975, *30*, 319–325.

Spirduso, W. W. Reaction and movement time as a function of age and physical activity level. *Journal of Gerontology*, 1975, *30*, 435–440.

Storandt, M., Wittels, I., and Botwinick, J. Predictors of a dimension of well-being in the relocated health aged. *Journal of Gerontology*, 1975, *30*(1), 97–102.

Terreberry, S. Household relocation: Resident's views. In E. Wolf and C. Lebeaux (Eds.), *Change and renewal in an urban community: Five case studies in Detroit, 1967*.

Wohlwill, J. F. Behavioral response and adaptation to environmental stimulation. In A. Damon (Ed.), *Physiological anthropology*, Cambridge, Mass.: Harvard University Press, 1974.

Woodruff, D. S. The usefulness of the life span approach for the psycho-physiology of aging. *The Gerontologist*, 1973, *13*, 467–472.

Woodruff, D. S., and Birren, J. E. (Eds.) *Aging*, New York: D. Van Nostrand Co., 1975.

Index